SEAWEED BIOMATERIALS

Edited by **Sabyasachi Maiti**

Seaweed Biomaterials

http://dx.doi.org/10.5772/intechopen.71251

Edited by Sabyasachi Maiti

Assistant to the Editor(s): Bibek Laha

Contributors

Hafiz Ansar Rasul Suleria, Sana Khalid, Munawar Abbas, Farhan Saeed, Huma Bader-Ul-Ain, Amir Husni, Dalia I. Sánchez-Machado, David R. Valenzuela-Rojo, Jaime López-Cervantes, Sabyasachi Maiti

Notice

Statements and opinions expressed in the chapters are these of the individual contributors and not necessarily those of the editors or publisher. No responsibility is accepted for the accuracy of information contained in the published chapters. The publisher assumes no responsibility for any damage or injury to persons or property arising out of the use of any materials, instructions, methods or ideas contained in the book.

First published in London, United Kingdom, 2018 by IntechOpen

IntechOpen is the global imprint of INTECHOPEN LIMITED, registered in England and Wales, registration number: 11086078, The Shard, 25th floor, 32 London Bridge Street

London, SE19SG – United Kingdom

Printed in Croatia

British Library Cataloguing-in-Publication Data

A catalogue record for this book is available from the British Library

Additional hard copies can be obtained from orders@intechopen.com

Seaweed Biomaterials, Edited by Sabyasachi Maiti

p. cm.

Print ISBN 978-1-78984-846-5

Online ISBN 978-1-78984-847-2

Meet the editor

Sabyasachi Maiti is an MPharm, PhD from Jadavpur University, Kolkata, India. He is currently working as an associate professor at the Department of Pharmacy, Indira Gandhi National Tribal University, Amarkantak, Madhya Pradesh, India. His research interest includes modification of natural polysaccharides and design of novel drug delivery systems. He has more than 50 research publications to his credit in various international journals of repute. He has contributed 20 book chapters and edited/co-edited books for various international publishers.

Contents

Chapter 1

Introductory Chapter: Seaweed-Derived Biomaterials

Sabyasachi Maiti

Additional information is available at the end of the chapter

http://dx.doi.org/10.5772/intechopen.82109

1. Introduction

Natural polymers and their application in the pharmaceutical industry are covered by the presence of synthetic polymers. Natural polysaccharides have gained popularity in the pharmaceutical, biomedical and food industry owing to their lucrative properties, namely biodegradability and biocompatibility and nontoxicity [1].

Polysaccharides can be obtained from a number of sources including seaweeds, bacteria, fungi and plants. When the polysaccharide is composed of only one kind of repeating monosaccharide, it is known as homo-polysaccharides, e.g., starch and cellulose. However, if the polysaccharide is composed of two or more different monomeric units, it is termed as hetero-polysaccharides, e.g., agar, alginate and carrageenan [2].

Further, the polysaccharides can be anionic, cationic and nonionic. Because they are extracted from natural resources, their composition and physicochemical properties vary considerably. Seaweed may belong to one of the several groups of multicellular algae: the red algae, green algae and brown algae. They are the rich sources of polysaccharides. Among the seaweed polysaccharides, alginates and carrageenan have been widely characterised and studied for the possible drug delivery application as well as their therapeutic potentials. In subsequent section of this introductory chapter, the therapeutic potential of these two polysaccharides is described.

2. Therapeutic potential of alginate

Alginate is derived from marine brown algae cell walls. Alginate is an anionic polysaccharide consisting of linear copolymer chain of (1-4)-linked β-D-mannuronic acid and α-L-guluronic acid in different arrangements of residues. Alginate is a natural, biodegradable and mucoadhesive polymer that does not produce toxicity in administration [3, 4]. This seaweed polysaccharide is most widely studied for the development of various kinds of drug delivery carriers. Its gel-forming ability in presence of divalent calcium ions has been extensively utilised for the fabrication of microparticles, nanoparticles and hydrogels in an attempt to achieve control drug release profiles. However, its bioactive potential has also been investigated and is described as follows.

Ueno and Oda [5] studied the effect of molecular weight and ratio of mannuronic acid/guluronic acid residues on TNF-α (tissue necrosis factor alpha) reducing activity in murine macrophage cell line. The alginates with molecular weight of 38,000 and the said ratio of 2.24 demonstrated potent TNF-α-inducing activity. They further noted that lyase depolymerised alginates were keen on inducing nitric oxide production from RAW264.7 cells compared to native alginate. However, both the polymers were equally effective with regards to their hydroxyl radical scavenging activities.

Its biocompatibility and similar gel texture and stiffness to that of the extracellular matrix have promoted alginate for its application in the field of tissue engineering and regeneration. Injectable alginate implants have been found to possess great potential in inhibiting the damaging events after myocardial infarction, leading to myocardial repair and tissue reconstruction [6]. Park et al. [7] depolymerised alginate following extraction from *Laminaria japonica*, and they found that the alginate lyase enzyme-treated 307 kDa alginate reduced the accumulation of lipid droplet and triglyceride in 3 T3-L1 preadipocytes in a dose-dependent manner and can be developed further for antiobesity treatment. The TNF-α secretion-inducing activity of alginate oligomers following enzymatic (lyase) hydrolysis of polyguluronate and polymannuronate differed in RAW264.7 cells and was dependent on the oligomer structures [8]. The encouraging preclinical results of cell-free alginate hydrogel implants in dogs with heart failure towards ventricle restoration have led to the initiation of clinical investigation for intramyocardial delivery of alginate implants in patients with acute myocardial infarction [9]. Houghton et al. [10] demonstrated that alginate in a bread vehicle can maintain its lipase inhibition properties despite cooking and digestion and therefore offers potential for the treatment of obesity. Wilcox et al. [11] reported that high-guluronic acid alginates isolated from *Laminaria hyperborea* seaweed could inhibit pancreatic lipase to a significantly higher degree than that obtained from *Lessonia nigrescens*. Therefore, lipase inhibition could be attributed to the variable structure of alginate. Their food use at high concentration could reduce the uptake of dietary triacylglycerol aiding in weight management. In another work by Chater et al. [12], the inhibitory effect of alginate on the gastro-oesophagus reflux aggressors trypsin and pepsin was investigated. The pepsin activity was reduced by up to 53.9% *in vitro*, although trypsin inhibition was trivial. The pepsin-reducing activity was correlated with the frequency of mannuronate residues in alginate. A significant proteolytic inhibition of dietary protein substrates was noted in the gastric phase of digestion, but not in the small intestinal phase.

3. Therapeutic potential of carrageenan

Carrageenan is a naturally occurring anionic sulphated linear polysaccharide extracted from certain red seaweed of the Rhodophyceae family [13]. Carrageenan consists of alternate units of D-galactose and 3,6-anhydro-galactose joined by α-1,3- and β-1,4-glycosidic linkage. Based on the amount and position of sulphate groups, carrageenan can be classified into lambda (λ), kappa (κ), iota (ι), nu (υ), mu (μ), theta (θ) and Ksi (ξ), all containing about 22–35% of sulphate groups [14]. Carrageenans have been extensively investigated for their anticoagulant, antiviral, cholesterol-lowering effects and immunomodulatory and antioxidant activity both *in vitro* and *in vivo* [15].

Sokolova et al. [16] reported that the antioxidant activity of carrageenans, i.e., their inhibitory effects on hydroxyl radicals and superoxide anion radicals, depends on the polysaccharide structure. Yuan et al. [17, 18] hydrolysed kappa-carrageenan to obtain oligosaccharides which were further sulphated, acetylated and phosphorylated. All the derivatives exhibited significant antioxidant activities; however, their antioxidant activity differs in different systems. The sulphated and acetylated derivatives scavenged superoxide radicals; the acetylated derivatives scavenged hydroxyl radicals, whereas the phosphorylated ones scavenged both hydroxyl radicals and DPPH radicals. The chemical modification of carrageenan oligosaccharides could enhance their antioxidant activity *in vitro*.

Carrageenan demonstrated encouraging antiviral activity against several animal viruses [19]. At a dose of 5 µg/ml, the destruction of the cell monolayer by herpes simplex virus type 1 (HSV-1) could be prevented. No evidence of cytotoxic effects was apparent up to a carrageenan concentration of 200 µg/ml. Girond et al. [20] reported a potent inhibitory effect of sulphated iota-, lambda- and kappa-carrageenans on the replication of hepatitis A virus (HAV) in the human hepatoma cell line, without any cytotoxic effects up to a strength of 200 µg/ml. Based on their selectivity indices, lota- and lambda-carrageenan was identified as promising candidates for chemotherapy of acute hepatitis A.

Carrageenans from *Gigartina acicularis* and *Eucheuma denticulatum* are more sulphated than those from *Kappaphycus cottonii*. At 0.75 mg/ml, no virucidal activity against HHV-1 or poliovirus was noticed. Their antiviral effect could be due to lower inhibition of the virus attachment and by the interference in the virus replication cycle [21].

Algal polysaccharides such as carrageenan are good sources of dietary fibre. Previous studies have shown that native carrageenan (κ) extracted from *Kappaphycus alvarezii* and commercial carrageenan have hypoglycaemic effects [22]. Panlasigui et al. [23] demonstrated that regular inclusion of carrageenan in the diet may result in reduced blood cholesterol and lipid levels in human subjects. Anderson et al. [24] reported that λ-carrageenans and κ-carrageenans from *Chondrus crispus* and *Polyides rotundus* possessed anticoagulant activity on intravenous injection in the rabbit. The differences in sulphate content between the carrageenans did not directly correlate with variable anticoagulant action and toxicity.

4. Conclusion

The literature reports suggested that seaweed polysaccharides possess potential pharmacological activities including antidiabetic, antiviral, antioxidant, anticoagulant, pepsin- and lipase-reducing and lipid-lowering activity. These polysaccharides either in their native form or modified form can be developed as therapeutic agents for the treatment of various diseases.

Further, they have been widely investigated as polymers for designing drug delivery carriers and tissue scaffolds. Overall, the exploration of these seaweed polysaccharides must be more focused to make them clinically useful as therapeutic agents.

Author details

Sabyasachi Maiti

Address all correspondence to: sabya245@rediffmail.com

Department of Pharmacy, Indira Gandhi National Tribal University, Madhya Pradesh, India

References

[1] Guo JH, Skinner GW, Harcum WW, Barnum PE. Pharmaceutical application of naturally occurring water soluble polymers. Pharmaceutical Science & Technology Today. 1998;**1**:254-261

[2] Prajapati VD, Maheriya PM, Jani GK, Solanki HK. Carrageenan: A natural seaweed polysaccharide and its applications. Carbohydrate Polymers. 2014;**105**:97-112

[3] Draget KI, Tylor C. Chemical, physical and biological properties of alginates and their biomedical implications. Food Hydrocolloids. 2011;**25**:251-256

[4] Downs EC, Robertson NE, Riss TL, Plunkett ML. Calcium alginate beads as a slow-release system for delivering angiogenic molecules *in vivo* and *in vitro*. Journal of Cellular Physiology. 1992;**152**:422-429

[5] Ueno M, Oda T. Biological activities of alginate. Advances in Food and Nutrition Research. 2014;**72**:95-112

[6] Ruvinov E, Cohen S. Alginate biomaterial for the treatment of myocardial infarction: Progress, translational strategies, and clinical outlook: From ocean algae to patient bedside. Advanced Drug Delivery Reviews. 2016;**15**:54-76

[7] Park M-J, Kim Y-H, Kim G-D, Nam S-W. Enzymatic production and adipocyte differentiation inhibition of low-molecular-weight-alginate. Journal of Life Sciences. 2015;**25**:1393-1398

[8] Iwamoto M, Kurachi M, Nakashima T, Kim D, Yamaguchi K, Oda T, et al. Structure-activity relationship of alginate oligosaccharides in the induction of cytokine production from RAW264.7 cells. FEBS Letters. 2005;**579**:4423-4429

[9] Lee RJ, Hinson A, Helgerson S, Bauernschmitt R, Sabbah HN. Polymer-based restoration of left ventricular mechanics. Cell Transplantation. 2013;**22**:529-533

[10] Houghton D, Wilcox MD, Chater PI, Brownlee IA, Seal CJ, Jeffrey P, et al. Biological activity of alginate and its effect on pancreatic lipase inhibition as a potential treatment for obesity. Food Hydrocolloids. 2015;**49**:18-24

[11]   Wilcox MD, Brownlee IA, Richardson JC, Dettmar PW, Pearson JP. The modulation of pancreatic lipase activity by alginates. Food Chemistry. 2014;**146**:479-484

[12]   Chater PI, Wilcox MD, Brownlee IA, Pearson JP. Alginate as a protease inhibitor in vitro and in a model gut system; selective inhibition of pepsin but not trypsin. Carbohydrate Polymers. 2015;**131**:142-151

[13]   Kirk RE, Othmer DF. In: Kroschwitz JI, HoweGrant M, editors. Encyclopedia of Chemical Technology. Vol. 4. New York: John Wiley & Sons; 1992. p. 942

[14]   Stanley NF. Carrageenans. In: McHugh DJ, editor. Production and Utilization of Products from Commercial Seaweeds. Australia: FAO Fisheries Technical Papers; 1987. ISBN: 9251026122

[15]   Pangestuti R, Kim SK. Biological activities of carrageenan. Advances in Food and Nutrition Research. 2014;**72**:113-124

[16]   Sokolova EV, Barabanova AO, Homenko VA, Solov'eva TF, Bogdanovich RN, Yermak IM. In vitro and ex vivo studies of antioxidant activity of carrageenans, sulfated polysac-charides from red algae. Bulletin of Experimental Biology and Medicine. 2011;**150**:426-428

[17]   Yuan H, Zhang W, Li X, Lü X, Li N, Gao X, et al. Preparation and in vitro antioxidant activity of kappa-carrageenan oligosaccharides and their oversulfated, acetylated, and phosphorylated derivatives. Carbohydrate Research. 2005;**340**:685-692

[18]   Yuan H, Song J, Zhang W, Li X, Li N, Gao X. Antioxidant activity and cytoprotective effect of kappa-carrageenan oligosaccharides and their different derivatives. Bioorganic & Medicinal Chemistry Letters. 2006;**16**:1329-1334

[19]   González ME, Alarcón B, Carrasco L. Polysaccharides as antiviral agents: Antiviral activity of carrageenan. Antimicrobial Agents and Chemotherapy. 1987;**31**:1388-1393

[20]   Girond S, Crance JM, Van Cuyck-Gandre H, Renaudet J, Deloince R. Antiviral activ-ity of carrageenan on hepatitis A virus replication in cell culture. Research in Virology. 1991;**142**:261-270

[21]   Jarbas M, Nathalie B, Boustie JB, Maryvonne A. Antiviral activity of carrageenans from marine red algae. Latin American Journal of Pharmacy. 2009;**28**:443-448

[22]   Suganya AM, Sanjivkumar M, Chandran MN, Palavesam A, Immanuel G. Pharmacological importance of sulphated polysaccharide carrageenan from red seaweed Kappaphycus alvarezii in comparison with commercial carrageenan. Biomedicine & Pharmacotherapy. 2016;**84**:1300-1312

[23]   Panlasigui LN, Baello OQ, Dimatangal JM, Dumelod BD. Blood cholesterol and lipid-lowering effects of carrageenan on human volunteers. Asia Pacific Journal of Clinical Nutrition. 2003;**12**:209-214

[24]   Anderson W, Duncan JGC, Harthill MJE. The anticoagulant activity of carrageenan. The Journal of Pharmacy and Pharmacology. 1965;**17**:647-654

Chapter 2

Therapeutic Potential of Seaweed Bioactive Compounds

Sana Khalid, Munawar Abbas, Farhan Saeed, Huma Bader-Ul-Ain and Hafiz Ansar Rasul Suleria

Additional information is available at the end of the chapter

http://dx.doi.org/10.5772/intechopen.74060

Abstract

Edible seaweeds are rich in bioactive compounds such as soluble dietary fibers, proteins, peptides, minerals, vitamins, polyunsaturated fatty acids and antioxidants. Previously, seaweeds were only used as gelling and thickening agents in the food or pharmaceutical industries, recent researches have revealed their potential as complementary medicine. The red, brown and green seaweeds have been shown to have therapeutic properties for health and disease management, such as anticancer, antiobesity, antidiabetic, antihypertensive, antihyperlipidemic, antioxidant, anticoagulant, anti-inflammatory, immunomodulatory, antiestrogenic, thyroid stimulating, neuroprotective, antiviral, antifungal, antibacterial and tissue healing properties. In proposed chapter, we discussed various active compounds include sulphated polysaccharides, phlorotannins, carotenoids (e.g. fucoxanthin), minerals, peptides and sulfolipids, with proven benefits against degenerative metabolic diseases. Moreover, therapeutic modes of action of these bioactive components and their reports are summarized in this chapter.

Keywords: seaweeds, marine, antioxidant, polysaccharides, bioactives

1. Introduction

Consumer interest has been increased from previous decades towards the health food and nutrition is the prime focus in formulating the food products. Algae are the organisms capable of providing bioactive compounds for producing novel medicinal and pharmaceutical substances. Algae are widely studied for human nutritional purpose and correspondingly utilized as functional foods [1]. Natural abundance, diverse origin and universal availability of algae makes it an essential source of biologically functional ingredients [2]. The term marine algae are generally referred as marine macroalgae or seaweeds [3]. Seaweeds are living resources found notably in littoral habitats or attached to rocks. They grow in shallow coastal waters as well as in deep sea areas up to a depth of 180 m. These macroscopic algae relatively occur in river mouth and saline waters. Seaweeds constitute the basis of the marine food chain and are subdivided in to three divisions, namely, brown algae, red algae and green algae [4].

Seaweeds, sometimes referred as edible marine algae, are regarded as good reservoir of compounds with numerous biological and biomedical activities and are most remarkably abundant in sulfated polysaccharides [5]. These have been studied in recent years to develop novel pharmaceuticals and potent bioactive substances [6, 7]. Edible macro algae have become a good source of food and alternative medicine in Asian countries [8] and in the western countries they are extraction specific and used for many industrial applications in food [9], cosmetics and pharmaceuticals [10]. The algal biotechnology industry is growing with an aquaculture division that produces large quantities of seaweeds, such as Laminaria, Gracilaria, and Spirulina. Additionally, the utilization phycocolloids derived from algae such as algin, agar, and carrageenan has developed into a well-established industry [11]. Cultivation of macroalgae now contributes to over 90% of the global seaweed demand, with the remainder being naturally harvested. Despite the growing worth of algae as a source of food ingredients, the industry has developed with only varying amounts of success and its biotechnological application are still under-exploited [12].

1.1. Types of seaweeds

At present, algae are divided in to four domains: Bacteria, Plantae, Chromista and Protozoa. All these vary greatly in morphology and sizes, which ranges from unicellular to multicellular microalgae or colony forming marine organisms such as macrophytes and seaweeds. Macroalgae are traditionally classified based on their characteristic forms and sizes, however the most commonly use feature in algal classification is the presence of specific pigments.

Marine algae due to their richness in bioactive compounds may exhibit antioxidant, anti-inflammatory, anticoagulant, antimicrobial, antiviral, antitumour and hypocholesterolemic activity [13]. Since seventeenth century marine macroalgae have long been used for biomedical purposes because of their potential phytochemical constituents and highly diverse nature. Algae can be classified into two groups based on their size: phytoplankton (microalgae) having 5000 different species and seaweed (macroalgae) with 6000 species [14].

Natural pigments determine the inherence of marine algae to one of the three algal divisions referred to as brown algae (*Phaeophyceae*), red algae (*Rhodophyceae*), and green algae (*Chlorophyceae*), respectively [3]. Brown colour of Phaeophyceae is due to the presence of pigment fucoxanthin. Red color of Rhodophyceae is often due to the dominance of phycoerythrin and phycocyanin pigments over the other pigments such as chlorophyll, carotene and xanthophylls. Green color of Chlorophyceae is due the presence of chlorophyll and related compounds in the same concentration as in higher plants. Some specific commercially important cultivated seaweeds and seaweed products include the brown seaweed *L. japonica*, from the brown seaweed *Undaria pinnatifida*, and Hizikia from *Hizikia fusiforme*. Biotechnological advances regarding macro algae cultivation include establishment of cell and tissue cultures that can biologically synthesize desired compounds, such as eicosanoids, on a large scale under a controlled environment [12].

1.2. Nutritional profile

Seaweeds have been recently emerged as a potential source of bioactive compounds with unique nutritional value and therapeutic activities, and it has become an important field of research in food science and technology [15, 16]. One of the main dietary differences between Eastern and Western hemispheres is the higher seafood consumption such as fish and marine algae [17]. Seaweeds are characterized as distinguished sources of various bioactive compounds with abundance in many minerals and could be utilized as novel functional foods which provide health benefit activities. Seaweed tissues are abundant in mineral elements such as iron, potassium, sulfur and iodine [18]. Depending upon seasonal conditions and the geographic area, macroalgae differs in the content of biochemical elements such as proteins, lipids, carbohydrates, vitamins and minerals [19]. Cell surface polysaccharides are responsible for high level of minerals and trace elements due to the retention of inorganic marine substances in marine algae [20]. Marine microalgae are considered as potential source of high quality proteins. *S. platensis* is considered as a prime source of bioactive proteins in marine environment. Compositional analysis of microalgae proteins clearly indicates that this high quality protein can be effectively used as direct supplements or could be used for formulation of other health products such as nutraceuticals [21].

Peptides with therapeutic potentials are referred as bioactive peptides and these peptides have potential applications in functional foods and nutraceuticals [22] for health improvement and better disease control. *Chlorella vulgaris*, *Spirulina platensis*, *Navicula incerta* and *Pavalova lutheri* are few potential algal species that could be used to extract biologically active peptides with significant therapeutic potentials is a widely studied marine microalga for extraction of bioactive peptides [23]. Mineral content of some seaweeds may account for up to 50%. Seaweeds species of kelp such as *Alaria esculenta* and *Chondrus crispus* are important vegetable sources of calcium [24]. The percentage of calcium can be as high as 7% of the dry weight and may be up to 34% in *Halimeda* sp. J.V. Lamouroux having calcified green segments [25]. *Laminaria digitata* is extensively used as a supplement for treatment of hypothyroidism and goiter [26]. The content weightage of calcium may reach 3% of the dry weight in macroalgae such as Fucus and Ascophyllum and up to 33.6% in calcified macroalgae such as *Phymatolithon calcareum* [27]. Therefore, consumption of seaweed could be beneficial to those at risk of calcium deficiency like pregnant females, teenagers and the elderly [28].

1.3. Bioactive compounds

Numerous metabolites extracted from marine algae possess biological activities. These bioactive compounds have been widely acknowledged because of their potential health benefits [3, 29]. Commercial bioactive compounds of algal origin include natural pigments (NPs), polyunsaturated fatty acids (PUFAs), lipids, proteins and polysaccharides [15, 16]. Some of these bioactive compounds with their sources are mentioned in **Table 1**. Natural variability in the content of bioactive molecules may be attributed to evolutionary relationships, ecological and chemical diversification but these should not be considered as limitations to commercialization [30]. Variation in the concentration of bioactive marine compounds of natural algal populations are influenced by environmental changes such as light, nutrients, contaminants, salinity, CO_2 availability, pH, temperature and biotic interactions [31].

Seaweeds	Bioactive compounds	References
Undaria pinnatifida	Fucoxanthin	[32]
Porphyra sp.	Phycoerythrobilin	[33]
Phaeophyceae	Sulfated fucoidans	[34]
Rhodophyceae	Sulfated galactans	[35]
Codium fragile	Xyloarabinogalactans	[36]
Codium cylindricum	Sulfated galactan	[37]
Sargassum thunbergii	Phlorotannins	[38]
Saccharina japonica	Fucoidans	[39]
Eisenia bicyclis	Phloroglucinol	[39]
Taonamaria atomaria	Stypoldione	[40]
Laurencia microcladia	Sesquiterpene elatol	[41]
Corallina pilulifera	Ethanolic extract	[41]
Schizymenia dubyi	Sulfated glucuronogalactan	[42]
Lobophora variegate	Fucans	[43]
Ecklonia cava	Phlorotannin 6,6'-bieckol	[44]
Porphyria dentate	Catechol, rutin and hesperidin	[45]

Table 1. Bioactive compounds from different seaweeds.

Functional materials of marine organisms occur in a wide variety and are enriched with polyunsaturated fatty acids, polysaccharides, pigments, minerals, vitamins, enzymes, phenolics and bioactive peptides [46]. Recently, the importance of algae as a source of structurally diverse bioactive compounds has been immensely emerged and research showed various biological activities of these compounds which are antioxidant, immunomodulation, anticoagulation and antiulcerogenic activities [47].

Seaweeds are the sole source of certain valuable phytochemicals, namely agar and carrageenan [48]. The richness of edible marine algae in sulfated polysaccharides (SPs) [49] as good sources of nutrients, span their uses from the food and pharmaceutical industries to biotechnology [5]. These anionic polysaccharide polymers are not only widespread in marine algae but also in mammals and invertebrates. Seaweeds are also the most significant sources of non-animal SPs and the chemical structure of these polymers vary according to the type of algae [50]. Major polysaccharides found in marine algae include fucoidan and laminarans found in brown algae, carrageenan in present red algae and ulvan in green algae [4].

Sulphated polysaccharides present in Rhodophyta are known as galactans which are composed of galactose or modified galactose units [51]. The class of Phaeophyta comprises of sulfated l-fucose units which are named as fucans. The polysaccharides found in Chlorophyta exhibit polydispersity among heteropolysaccharides together with traces of homopolysaccharides [50]. Carrageenan may also show anticoagulant activity [52], antiviral activity [53], and antitumor activity [54]. Marine red algae primarily contain an agaran type polysaccharide, which was separated from *Grateloupia filicina* and was investigated for its antiangiogenic activity.

Fucoidan is a highly complex sulfated polysaccharide found in marine brown algae is also present in microorganisms, plants and animals [44]. Fucoidan have been shown to exhibit antiviral and anti-inflammatory affect. Anti-metastatic effects of fucoidan obtained from *Fucus vesiculosus*, have been described. Fucoidan could also be reflected as a potential therapeutic agent against the metastasized invasive human lung cancer cells. Phloroglucinol bioactives acquired from marine seaweeds have chemical diversity and are much studied for their remarkably beneficial biological actions.

Seaweeds have been majorly studied for their biologically active polyphenolic derivatives called phlorotannins [43]. Marine brown algae (*Phaeophyta*) accumulate a variety of phloroglucinol based polyphenols, as phlorotannins [47]. Among marine brown algae, *Ecklonia cava, Ecklonia stolonifera, Ecklonia kurome, Eisenia bicyclis, Sargassum thunbergii, Hizikia fusiformis, Undaria pinnatifida* and *Laminaria japonica* have been reported to exhibit health beneficial activities because phlorotannins. Due to the various biological activities of phlorotannins, marine brown algae are known to be a rich source of healthy food [55]. *Undaria pinnatifida* contain 5–10% fucoxanthin and it is one of the most well-known edible seaweed in Japan. Health benefits of fucoxanthin include anticancer effect and it is reported that neoxanthin and fucoxanthin cause a significant reduction in growth of prostate cancer cells. Anti-obesity activity and anti-inflammatory activity was also demonstrated [56]. Fucoxanthin is other major biofunctional pigment of brown seaweeds and has been found in high concentration in various edible seaweeds including *U. Pinnatifida* [57].

2. Remedial activities

2.1. Antioxidant activity

Antioxidants may affect the human health in a positive way as they can protect the human body against damage by Reactive oxygen species, which attack and impair macromolecules such as DNA, proteins and lipids lead to many health disorders such as diabetes, aging, cancer and other neurodegenerative diseases [58]. Recently, marine flora and fauna gain considerable interest as natural sources for the development of antioxidants in the food and pharmaceutical industry. Marine algae represent one of the richest sources of natural antioxidants among marine resources [59]. Antioxidant activity of marine derived bioactive peptides has been determined through radical scavenging activities which have been detected by electron spin resonance spectroscopy method as well as intra cellular free radical scavenging assays. The peptide chain contains hydrophobic amino acids which contribute towards their potential antioxidant activity [60, 61].

Marine algae have various classes of natural polysaccharides including fucoxanthin, phycoerythrobilin, chlorophyll-a and their derivatives show potent antioxidant activity. Cho et al. [62] suggested that strong antioxidant activity of the *Enteromorpha prolifera* was caused by chlorophyll-a derivatives, pheophorbidea, rather than phenolic compounds. The antioxidant activity is due to the specific scavenging of oxygen or radicals [63] formed during peroxidation or metal-chelating ability [64]. Yan et al. [32] discovered that fucoxanthin show strong radical scavenging activity [65] which isolated fucoxanthin from *Undaria pinnatifida* and prepared two fucoxanthin metabolites, fucoxanthinol and halocyn-thiaxanthin. Hence, fucoxanthin serves as substitute for synthetic antioxidants in nutraceuticals and pharmaceuticals. Cytoprotective effect of fucoxanthin has been observed in vitro against ROS formation induced by H_2O_2. Two hydroxyl groups are present in the ring structure of fucoxanthin, which are responsible for the inhibition of ROS formation. Several studies supported the fact that number of hydroxyl groups on the ring structure of fucoxanthin causes the effects of ROS suppression [66]. Recently, Yabuta et al. [33] demonstrated antioxidant activity of phycoerythrobilin derived from *Porphyra* sp.

NPs are useful effective bioactive substances in search for effective, non-toxic substances with potential antioxidant activity. NPs are distributed in large quantities in marine algae and could be used as a rich source of natural antioxidants with potential application in the food industry as well as cosmetic and pharmaceutical areas [3].

2.2. Anti-coagulant activity

Whenever an abnormal vascular condition occurs, blood coagulation begins to stop the flow of blood though the injured vessel wall and exposure to non-endothelial surfaces at sites of vascular injury occur. Blood coagulation is processed by coagulation factors. The blood coagulation can be prolonged or stopped when endogenous or exogenous anticoagulants interfere with these coagulation factors [67]. The anticoagulants derived from marine bioactive peptides have been extensively reported, but they have also been isolated from marine organisms such as marine echiuroid worm [68]. The anticoagulant activity of the bioactive peptides has been determined by prolongation of prothrombin time, thrombin time and activated partial thromboplastin time assays and the activity was compared with the standard commercial anticoagulant heparin. The normal clotting time of anticoagulant peptide isolated from marine echiuroid worm have been significantly prolonged [69].

Sulphated polysaccharides derived from marine brown algae are alternate sources for manufacturing of novel anticoagulants [37]. Anticoagulant activity is the most extensively studied property of sulphated polysaccharides and have been reviewed previously [70]. Two types of SPs have been recognized with high anticoagulant activity. Marine red algae produce sulfated galactans also known as carrageenan, [35] and marine brown algae produce sulfated fucoidans [34]. There are very few reports of anticoagulant SPs reported from marine green algae. Jurd et al. [36] found that the anticoagulant active SPs from *Codium fragile* contain xyloarabinogalactans. *Codium cylindricum* also contain a sulfated galactan with anticoagulant activity. Additionally, Maeda et al. [71] have revealed that the anticoagulant SPs from *Monostroma nitidum* yield a six fold higher activity as compared to heparin. Marine brown algae extracts demonstrate higher anticoagulant activity than red and green algae extracts [34]. The presence of sulfate functional groups in SPs can increase both the specific as well as nonspecific binding to a wide-range of biologically active proteins. Anticoagulant activity of sulfated galactans depends on the sulfate content, the sulfation position of the structure, and nature of the sugar residue in SPs [72]. High molecular weight carrageenans having high sulfate content show higher anticoagulant activity in comparison to low molecular weight carrageenans having low sulfate content of SPs [73].

Low molecular weight and unfractionated heparins are the only sulfated polysaccharides currently used as anticoagulant drugs. Seaweed derived SPs possess anticoagulant activity similar to or higher than the heparin [50]. In the pharmaceutical industry, SPs derived from seaweeds are the promising bioactive agents to be used as anticoagulant agents. Phlorotannins derived from *Sargassum thunbergii* are potential anticoagulants in vitro and in vivo. These phlorotannins from *S. thunbergii* had a significant effect on the prolongation of prothrombin time, thrombin time and activated partial thromboplastin time. In addition, phloroglucinol can be established as a novel anticoagulant in pharmaceutical industry [38].

2.3. Anti-cancer activity

Marine algae produce a range of diverse anti-cancer phytochemicals. Based on epidemiological data, the protective effect of edible seaweeds has been established against mammary, skin and intestinal carcinogenesis [74]. The bioactive substances can kill cancerous cells by inducing apoptosis or they may affect cell signaling by the activation of cell signaling enzymes of protein kinase-c family of brown algae seaweeds [75]. *Laminaria, Gelidiumamansii* and *Porphyratenera* exhibit dose-dependent inhibition of growth in mutated human gastric and colon cells [76] and also cancer cells of mammary glands. Brown seaweeds such as *Laminaria* are edible as a functional food, and it is well known for reducing the incidence of breast cancer in Japan to about one sixth as that of the rate reported for American women. *Laminaria japonica* and *Sargassum muticum* species are widely used as components of conventional herbal medicines for the treatment of cancer in china [77].

Most remarkable compounds found naturally in the brown seaweeds are the fucoidans, glucans and some other secondary [75] metabolites. Most of these compounds are listed in (**Table 2**). These compounds are capable of producing anticancer activity. Fucoidans from *Saccharina japonica* and *Undaria pinnatifida* dose-dependently inhibit proliferation and colony formation in both breast cancer and melanoma cell. This proves that the use of sulfated polysaccharides from both above mentioned brown seaweeds are potential ingredients for cancer treatment. Low molecular weight fucoidan isolated from *Ascophyllumnodosum* selectively inhibits the invasion of breast cancer cells by a mechanism of blocking the accession of these cancerous cells in the extracellular matrix and it also inhibits the invasive colon adenocarcinoma cells [78].

Phloroglucinol and its essential polymers which are eckol, dieckol, phlorofucofuroeckol A, and 8,8′-bieckol isolated from the brown alga *Eisenia bicyclis* show significant anticancer activity [39]. The extract of the brown alga *Taonamaria atomaria* contains a compound stypoldione, an in-vitro inhibitor of microtubule polymerization, exhibits anticancer activity [40]. Red algae contain abundant concentration of secondary metabolites and their halogenated derivatives. *Laurencia microcladia* produces a sesquiterpene elatol exhibit antitumour activity. Elatol exhibits cytotoxicity by inducing cell cycle arrest leading cells to apoptosis. *Corallina pilulifera* is a calcareous red alga whose ethanolic extract show anti-proliferative activity on human cervical adenocarcinoma cells. *Acanthospora spicifer*, another red seaweed, exhibits tumouricidal activity against Ehrlich's ascites carcinoma cells. This is due to decrease in tumour volume

Health claims	Bioactive compounds	References
Antioxidant activity	• Fucoxanthin • Phycoerythrobilin • Chlorophyll-a and derivatives	[76]

Anti-coagulant activity	• Sulphated polysaccharides
	   ○ Galactans/Carrageenan
	   ○ Fucoidans
	   ○ Heparins
	   ○ Phlorotannins
	   ○ Phloroglucinol
		[37]

Anti-cancer activity	• Fucoidans • Glucans • Phloroglucinol • Stypoldione • Sesquiterpene elatol • Carotene • Lutien	[39]
Anti-viral/Anti-HIV activity	• Sulfated Glucuronogalactan • Sulphated galactans • Sulphated fucans • Carrageenan	[54]
Cardiovascular protection	• Carotenoids • Sterols • Cardiac glycosides • Eicosapentaenoic acid • Docosa-hexaenoic acid	[81]
Anti-inflammatory activity	• Marine terpenes • Bioactive peptides • Sulfated Polysaccharides • Fucoidan • Ascophyllan • Algal polyphenols • Phlorotannins • Terpenes and steroids • Alkaloids • Commercially produced microalgal PUFAs	[82]

Table 2. Bioactive compounds from seaweeds with their health promoting functions.

and viable cell counts and increase in the mean survival time [41]. *Porphyratenera,* a red alga, has been extensively reported for its high anti-carcinogenic effect [79]. Chlorophyll-related compounds, carotene and lutien isolated from algae exhibit strong anti-mutagenic activity in vitro as well as in vivo [80].

Various anticancer pathways are involved to accomplish the process of tumour cell death. Major pathways are anti-oxidation and immune stimulation, and apoptosis of cancerous cells. Tumors are in a 'pro-oxidant' state generating more free radicals. These free radicals usually accompanied by lack of DNA repair mechanisms. Reactive oxygen species are main sources of oxidative stress in cells, damaging DNA, proteins and lipids. Anti-oxidants cause inhibition of the growth of cancer cells through varied mechanisms. The most common is activation of apoptosis by antioxidant species and inhibition the process of tumour progression [64].

Apoptosis is a process of programmed cell death triggered by various extrinsic or intrinsic stimuli in unfavourable situations. The protein p53 and caspase-cascade signaling system are prime factors for promoting apoptosis [83]. Caspases belong to the interleukin 1β converting enzyme family of proteases. The process of apoptosis has three stages, namely activation, execution and cell deletion. All these stages are interlinked by caspases [84]. Tumour suppressor protein p53 triggers the apoptosis and induces cell growth arrest. The prevention of cancer is highly dependent on p53 for controlling the proliferation of cells with damaged DNA or with a potential for neoplastic transformation. Algae is a source of many phytochemicals which cause apoptosis. *Spirulina* and *Aphanizomenon fos-aquae* are two most common edible cyanobacteria [85]. Both contain phycocyanin, which is capable of showing apoptosis in the chronic myeloid leukaemia cells. Enzymatic extraction of alga, *Ecklonia cava* together with its polysaccharides and polyphenolics, displays tremendous anti-proliferative activity against cancer cell line [75].

The apoptosis is executed immunostimulation with two pathways, the NK cell and Fas receptor mediated pathways. The Fas receptor molecule plays an important role in the immune system, which allows the removal of auto-antibodies and the elimination of virally infected tumourigenic cells. Immune defence mechanisms do kill any abnormal cells including cancer. Polysaccharides are associated with biological activities of several microalgal species. Polysaccharide complexes from *Chlorella pyrenoidosa* contain glucose and any combination of mannose, galactose, arabinose, and rhamnose. The complexes *N*-acetylglucosamide and *N*-acetylgalactosamine have immune stimulating properties and can inhibit the proliferation of pathogenic microbes such as *Listeria monocytogene s*and *Candida albicans* [75] *Enteromorpha compressa,* produces a range of bioactive compounds which are proved to be useful in the treatment of cancer and inflammation [86]. Malyngamides isolated from *Lyngbya majuscule* have immunosuppressant properties and is also cytotoxic [87].

2.4. Anti-viral/Anti- HIV activity

Acquired immunodeficiency syndrome AIDS is a disease caused by human immunodeficiency virus (HIV-1) [88]. Marine algae derived SPs can inhibit replication of enveloped viruses such as herpes virus, togavirus, arenavirus, rhabdovirus, and orthopoxvirus families.

These sulaphated polysaccharides have great potential for the development of novel anti-HIV therapeutics. Marine algae possess significant quantities of complex structural SPs that inhibit the HIV. The chemical structure, constituent sugars, stereochemistry, degree of sulfation, molecular weight and conformation are affected the antiviral activity of algal SPs [89].

SPs from red algae also exhibit significant HIV-1 inhibitory activity. Anti-HIV activity of *Schizymenia dubyi* is due to sulfated glucuronogalactan. This polysaccharide causes inhibition of virus-host cell attachment in vitro. A mechanism which occur mainly during initial step of HIV infection [42]. Additionally, antiretroviral activity of sulfated galactans from *Grateloupia filicina* and *Grateloupia longifolia* was examined with a primary isolate of HIV-1 and human peripheral blood mononuclear cells [54]. Sulfated fucans from the brown seaweed *F. vesiculosus, Lobophora variegata, Dictyota mertensiian* and *Spatoglossum schroederi* were reported to inhibit HIV reverse transcriptase [90]. Human papilloma virus is the cause of cervical cancer due to infection in female genital tract. Therefore, HPV Infection control has acquired great attention from scientific studies [91]. Natural bioactive compounds and their derivatives are potential source for the manufacture of functional foods as novel anti-HPV therapeutics with fewer side effects, more effective and cost effective. Marine algae contain substantial quantities of complex structural SPs which are potent inhibitors of wide variety of viruses, such as papilloma virus [92].

Carrageenan has been shown to demonstrate anti-HPV activity in vitro [92]. Carrageenan inhibits HPV 3-fold higher in magnitude than heparin, a highly effective model for HPV. Carrageenan acts mainly by preventing the binding of HPV virions to cells and blocks HPV infection through a post attachment heparin sulfate-independent effect. This mechanism is consistent by the fact that carrageenan closely resembles heparin sulfate, which is recognized as HPV-cell attachment factor. Moreover, antigen-specific immune responses and antitumor effects of carrageenan were remarkable [93]. Carrageenan are the promising candidates for production of new therapeutic agents for HPV by being a part of food additives. There are numerous advantages of carrageenan over other classes of antiviral agents, such as reasonably low production costs, novel modes of action, broad spectrum, low cytotoxicity and safety [92].

2.5. Anti-cardiovascular disease activity

Dyslipidemia is a main cardiovascular risk factor for coronary heart disease incidence and mortality. Lipid disorders can accelerate the atherosclerosis process and result could be chronic heart failure. Nutraceuticals are effectively able to reduce the atherosclerosis process and coronary heart disease progression. Carotenoids are produced by seaweeds, plants and microorganisms. These fat soluble are the fundamental component of Mediterranean foods, are well known to reduce the incidence and frequency of cardiovascular events, perhaps by means of their antioxidant action on free radicals or by anti-inflammatory action on lipoxygenase enzyme activity [94].

Cell membranes contain sterols as important structural components, and some of them are cardiac glycosides used therapeutically in the treatment of cardiac failure and atrial arrhythmias. The positive effect of eicosapentaenoic acid and docosa-hexaenoic acid on human health has been reported as far as cardiovascular system. Enrichment of foods with EPA/DHA show cardio protective effects. EPA and DHA may exert their cardio protective functions, namely influencing plasmatic triacylglycerol (TAG) and cholesterol levels, and modulation of the chronic inflammation in the vascular wall [95].

2.6. Anti-inflammatory activity

Inflammation underlies a mass of enormous malignancies such as asthma, myocardial ischemia, allergies, arthritis, atherosclerosis and cancer. Inflammation is a complex biological process and occurs in response to harmful stimuli such as presence of pathogens in vascular tissues or injury. Inflammation normally acts as a defense mechanism, and its deregulation is associated with a multitude of diseases. Chronic and acute inflammation is a physiological process mediated by the activation of immune cells such as mononuclear phagocytic cells and macrophages [96]. The mechanism of inflammation is controlled by endogenous chemical mediators such as vasoactive amines, platelet activating factor, cytokines, bradykinin, fibrin, complement component, eicosanoids, nitric oxide and reactive oxygen species. These inflammatory mediators play a pivotal role in controlling various steps of inflammation. Marine algae produce a diverse array of secondary metabolites which play a pivotal role as inhibitors of inflammation [97].

Marine algae produce a combination of metabolites which are implicated in large number of diseases because of anti-inflammatory and antioxidant properties, with high commercial utilization. These compounds include fatty acids, marine terpenes, bioactive peptides, polysaccharides and their structures ranges from aliphatic molecules with a linear chain to complex polycyclic entities. Marine sulfated PS exhibit anti-coagulant, anti-inflammatory, anti-viral and anti-tumor activities and are important in pharmaceutical industries [82].

These algal compounds usually possess immune-modulatory activities which potentially instigate the immune system activities to alleviate undesirable responses such as inflammation. Sulfated polysaccharides may target numerous pathways in the immune and inflammatory systems. They can affect disease pathophysiology and outcome, including tumour development and septic shock. Fucoidan possess extensive of biological activities which include anti-inflammatory and anti-oxidative effects. Research revealed that the mechanism behind anti-inflammatory effect of fucoidan is due to its capability to interact with an adhesion molecule selectin on the seaweed cell membrane [98]. Fucoidan show anti-oxidative effect by inhibiting the synthesis and release of reactive oxygen radicals as well as its clearance. Park et al. [43] studied the cellular and molecular mechanism underlying the anti-inflammatory properties of fucoidan.

According to research, ascophyllan is a discrete sulfated polysaccharide isolated from fucoidan with significant biological activity. *Lobophora variegate* is a brown marine alga, which possess a high content of fucans exhibit reduced anti-inflammatory process *in vivo*. Two sulfated PS from *Laminaria saccharina*, a brown seaweed, utilized for the treatment of inflammation. Sulfated polysaccharides of the seaweed *L. variegate* exhibit antioxidant power and anti-inflammatory activity against zymosan induced arthritis [99]. Sulfated PS 'sacran' is also of marine algal origin. A sulfated polysaccharide isolated from *Aphanothece sacrum* exerts an epicutaneous effect on 2,4,6-tncb (picryl chloride) induced allergic dermatitis *in vivo* by improving functions of skin barriers and by decreasing the pro-inflammatory cytokine production [100].

Algal polyphenols and phlorotannins have numerous biological properties besides their strong antioxidant properties. Phlorotannins are the main bioactive compounds found in marine algae. Yang *et al.* [44] proposed the underlying anti-inflammatory mechanism of the phlorotannin 6,6'-bieckol, an active component isolated from brown seaweed *Ecklonia cava.* These findings suggest that the anti-inflammatory properties of this compound are related to the inhibition of cyclooxygenase-2 and pro-inflammatory cytokines (TNF-$\alpha$and IL-6).

Porphyria dentate is a red edible seaweed and its use in treatment of inflammatory diseases was the long lasting tradition globally. Crude extract of *P. dentate* contain phenolic compounds such as catechol, rutin and hesperidin [45]. Researcher demonstrated that the therapeutic applications of c-phycocyanin obtained from blue-green algae Spirulina *platensis* that significantly suppress the activation of LPS-induced nitrite and iNOS protein expression, accompanied by an attenuation of TNF-α formation. Marine red algae are the source of anti-inflammatory cyclic dipeptides and diketopiperazine [101] Terpenes and steroids are the classes of anti-inflammatory compounds found ubiquitously in marine algae. Heo et al. [102] evaluated the potential of fucoxanthin to produce anti-inflammatory effect via inhibition of NO production and reduced Prostaglandin-E2 production. Further investigations indicated the suppression of iNOS and COX-2 mRNA expressions by fucoxanthin in LPS-stimulate macrophage cells. By the addition of fucoxanthin in a dose-dependent manner, the release of cytokines TNF-α, IL-1β and IL-6 were also reduced [102].

Alkaloids occur rarely in marine algae, alkaloids isolated from marine algae have been shown to possess anti-inflammatory properties [103]. Algal fatty acids are either saturated or unsaturated with reported bioactivity. Commercially produced microalgal PUFAs are of particular interest because they lead the human body to more anti-inflammatory environment. Various benefits accrued from docosahexaenoic acid and palmitoleic acid are the reduction in the incidence of certain heart diseases and oleic acid retain antioxidant capacity [16].

3. Conclusion

Seaweeds are a valuable source of bioactive compounds and could be introduced for the preparation of novel functional ingredients in food and also a good approach for the treatment or prevention of chronic diseases. Recently, much att`ention has been paid by the consumers toward natural bioactive compounds as functional ingredients in foods, and hence, it can be suggested that, seaweeds are an alternative source for synthetic ingredients that can contribute to consumer's well-being, by being a part of new functional foods and pharmaceuticals. Furthermore, the wide ranges of biological activities associated with marine algae-derived bioactive compounds have potential to expand its health beneficial value in food, and pharmaceutical industries.

Conflict of interest

Authors declare no potential conflict of interest.

Author details

Sana Khalid[1], Munawar Abbas[2], Farhan Saeed[2], Huma Bader-Ul-Ain[2] and
Hafiz Ansar Rasul Suleria[3,4,5*]

*Address all correspondence to: hafiz.suleria@uqconnect.edu.au

1 Department of Pharmaceutical Sciences, Government College University, Faisalabad,
Pakistan

2 Institute of Home and Food Sciences, Government College University, Faisalabad,
Pakistan

3 UQ Diamantina Institute, Translational Research Institute, Faculty of Medicine,
The University of Queensland, Brisbane, Australia

4 Department of Food, Nutrition, Dietetics and Health, Kansas State University,
Manhattan, KS, USA

5 Centre for Chemistry and Biotechnology, School of Life and Environmental Sciences,
Deakin University, Waurn Ponds, Australia

References

[1] Holdt SLV, Kraan S. Bioactive compounds in seaweed: Functional food applications and legislation. Journal of Applied Phycology. 2011;**23**(3):543-597

[2] Domanguez H. Algae as a source of biologically active ingredients for the formulation of functional foods and nutraceuticals. In: Functional Ingredients from Algae for Foods and Nutraceuticals. Cambridge: WoodHead; 2013. pp. 1-19

[3] Pangestuti R, Kim SK. Biological activities and health benefit effects of natural pigments derived from marine algae. Journal of Functional Foods. 2011;**3**(4):255-266

[4] Mayakrishnan V, Kannappan P, et al. Cardioprotective activity of polysaccharides derived from marine algae: An overview. Trends in Food Science & Technology. 2013;**30**(2):98-104

[5] Renn D. Biotechnology and the red seaweed polysaccharide industry: Status, needs and prospects. Trends in Biotechnology. 1997;**15**(1):9-14

[6] Suleria HAR. Marine Processing Waste-In Search of Bioactive Molecules. Natural Products Chemistry and Research. 2016;**4**:e118

[7] Li B, Lu F, et al. Fucoidan: Structure and bioactivity. Molecules. 2008;**13**(8):1671-1695

[8] Muhammad SA, Muhammad J, et al. Metabolites of marine algae collected from Karachi-coasts of Arabian Sea. Natural Product Sciences. 2000;**6**(2):61-65

[9] Suleria HAR, Gobe G, Masci P, Osborne SA. Marine bioactive compounds and health promoting perspectives; innovation pathways for drug discovery. Trends in Food Science & Technology. 2016;**50**:44-55

[10] Gamez-Ordaaez E, Jimanez-Escrig A, et al. Dietary fibre and physicochemical properties of several edible seaweeds from the northwestern Spanish coast. Food Research International. 2010;**43**(9):2289-2294

[11] Tseng C. Algal biotechnology industries and research activities in China. Journal of Applied Phycology. 2001;**13**(4):375-380

[12] Rasmussen RS, Morrissey MT. Marine biotechnology for production of food ingredients. Advances in Food and Nutrition Research. 2007;**52**:237-292

[13] Frestedt JL, Kuskowski MA, et al. A natural seaweed derived mineral supplement (Aquamin F) for knee osteoarthritis: A randomised, placebo controlled pilot study. Nutrition Journal. 2009;**8**(1):7

[14] Ravikumar S, Kathiresan K. Influence of tannins, amino acids and sugars on fungi of marine halophytes. Mahasagar. 1993;**26**(1):21-25

[15] Plaza M, Cifuentes A, et al. In the search of new functional food ingredients from algae. Trends in Food Science & Technology. 2008;**19**(1):31-39

[16] Plaza M, Herrero M, et al. Innovative natural functional ingredients from microalgae. Journal of Agricultural and Food Chemistry. 2009;**57**(16):7159-7170

[17] Terry PD, Rohan TE, et al. Intakes of fish and marine fatty acids and the risks of cancers of the breast and prostate and of other hormone-related cancers: A review of the epidemiologic evidence. The American Journal of Clinical Nutrition. 2003;**77**(3):532-543

[18] Kraan S. Pigments and minor compounds in algae. In: Domanguez H, editor. Functional Ingredients from Algae for Foods and Nutraceuticals. Cambridge: WoodHead; 2013. pp. 205-251

[19] Zubia M, Payri C, et al. Alginate, mannitol, phenolic compounds and biological activities of two range-extending brown algae, *Sargassum mangarevense* and *Turbinaria ornata* (Phaeophyta: Fucales), from Tahiti (French Polynesia). Journal of Applied Phycology. 2008;**20**(6):1033-1043

[20] Mabeau S, Jl F. Seaweed in food products: Biochemical and nutritional aspects. Trends in Food Science & Technology. 1993;**4**(4):103-107

[21] Brown MR. Nutritional value and use of microalgae in aquaculture. In: Avances en Nutrician Acuacola VI. Memorias del VI Simposium Internacional de Nutrician Acuacola. 2002;**3**:281-292

[22] Suleria HAR, Masci P, Gobe G, Osborne S. Current and potential uses of bioactive molecules from marine processing waste. Journal of the Science of Food and Agriculture. 2016;**96**(4):1064-1067

[23] Dewapriya P, Kim SK. Marine microorganisms: An emerging avenue in modern nutraceuticals and functional foods. Food Research International. 2014;**56**:115-125

[24] Nisizawa K. Seaweeds Kaiso: Bountiful Harvest from the Seas. Japan Seaweed Association. Shimane; 2002. pp. 44-50

[25] Arasaki S, Arasaki T. Vegetables from the Sea. Vol. 96. Tokyo: Japan Publ. Inc.; 1983. pp. 251-223

[26] Massig K. Iodine-induced toxic effects due to seaweed consumption. In: Comprehensive Handbook of Iodine. Amsterdam: Elsevier; 2009. pp. 897-908

[27] Blunden G, Campbell SA, et al. Chemical and physical characterization of calcified red algal deposits known as maarl. Journal of Applied Phycology. 1997;9(1):11-17

[28] Burtin P. Nutritional value of seaweeds. Electronic Journal of Environmental, Agricultural and Food Chemistry. 2003;2(4):498-503

[29] Suleria HAR, Osborne S, Masci P, Gobe G. Marine-based nutraceuticals: An innovative trend in the food and supplement industries. Marine Drugs. 2015;13:6336-6351

[30] Suleria HR, Butt MS, Anjum FM, Arshad M, Khalid N. Aqueous garlic extract mitigate hypercholesterolemia and hyperglycemia; rabbit experimental modelling. Annals of Nutrition and Metabolism. 2013;63:271

[31] Stengel DB, Connan SN, et al. Algal chemodiversity and bioactivity: Sources of natural variability and implications for commercial application. Biotechnology Advances. 2011;29(5):483-501

[32] Yan X, Chuda Y, et al. Fucoxanthin as the major antioxidant in *Hijikia fusiformis*, a common edible seaweed. Bioscience, Biotechnology, and Biochemistry. 1999;63(3):605-607

[33] Yabuta Y, Fujimura H, et al. Antioxidant activity of the phycoerythrobilin compound formed from a dried Korean purple laver (*Porphyra* sp.) during in vitro digestion. Food Science and Technology Research. 2010;16(4):347-352

[34] Chevolot L, Foucault A, et al. Further data on the structure of brown seaweed fucans: Relationships with anticoagulant activity. Carbohydrate Research. 1999;319(1):154-165

[35] Kolender AA, Pujol CA, et al. The system of sulfated -linked D-mannans from the red seaweed *Nothogenia fastigiata*: Structures, antiherpetic and anticoagulant properties. Carbohydrate Research. 1997;304(1):53-60

[36] Jurd KM, Rogers DJ, et al. Anticoagulant properties of sulphated polysaccharides and a proteoglycan from *Codium fragile* ssp. atlanticum. Journal of Applied Phycology. 1995;7(4):339-345

[37] Matsubara K. Recent advances in marine algal anticoagulants. Current Medicinal Chemistry. Cardiovascular and Hematological Agents. 2004;2(1):13-19

[38] Bae JS. Antithrombotic and profibrinolytic activities of phloroglucinol. Food and Chemical Toxicology. 2011;49(7):1572-1577

[39] Shibata T, Fujimoto K, et al. Inhibitory activity of brown algal phlorotannins against hyaluronidase. International Journal of Food Science and Technology. 2002;37(6):703-709

[40] Mayer AM, Lehmann VK. Marine pharmacology in 1998: Marine compounds with antibacterial, anticoagulant, antifungal, antiinflammatory, anthelmintic, antiplatelet, antiprotozoal,

and antiviral activities; with actions on the cardiovascular, endocrine, immune, and nervous systems; and other miscellaneous mechanisms of action. The Pharmacologist. 2000;**42**(2):62-69

[41] Vasanthi H, Rajamanickam G, et al. Tumoricidal effect of the red algae *Acanthophora spicifera* on Ehrlich ascites carcinoma in mice. Seaweed Research and Utilization. 2004:217-224

[42] Bourgougnon N, Lahaye M, et al. Annual variation in composition and in vitro anti-HIV-1 activity of the sulfated glucuronogalactan from *Schizymenia dubyi* (Rhodophyta, Gigartinales). Journal of Applied Phycology. 1996;**8**(2):155-161

[43] Park HY, Han MH, et al. Anti-inflammatory effects of fucoidan through inhibition of NF-IB, MAPK and Akt activation in lipopolysaccharide-induced BV2 microglia cells. Food and Chemical Toxicology. 2011;**49**(8):1745-1752

[44] Yang L, Zhang LM. Chemical structural and chain conformational characterization of some bioactive polysaccharides isolated from natural sources. Carbohydrate Polymers. 2009;**76**(3):349-361

[45] Kaza-owska K, Hsu T, et al. Anti-inflammatory properties of phenolic compounds and crude extract from *Porphyra dentata*. Journal of Ethnopharmacology. 2010;**128**(1):123-130

[46] Shahidi F, Alasalvar C. Marine oils and other marine nutraceuticals. In: Handbook of Seafood Quality, Safety and Health Applications. Chichestar: Blackwell; 2011. pp. 444-454

[47] Barrow C, Shahidi F. Marine Nutraceuticals and Functional Foods. New York. CRC Press; 2007

[48] Ismail A, Hong TS. Antioxidant activity of selected commercial seaweeds. Malaysian Journal of Nutrition. 2002;**8**(2):167-177

[49] Suleria HAR, Masci PP, Zhao KN, Addepalli R, Chen W, Osborne SA, Gobe GC. Anti-coagulant and anti-thrombotic properties of blacklip abalone (*Haliotis rubra*): In vitro and animal studies. Marine Drugs. 2017;**15**(8):240

[50] Costa L, Fidelis G, et al. Biological activities of sulfated polysaccharides from tropical seaweeds. Biomedicine & Pharmacotherapy. 2010;**64**(1):21-28

[51] Mayer AM, Rodraguez AD, et al. Marine pharmacology in 2007: Marine compounds with antibacterial, anticoagulant, antifungal, anti-inflammatory, antimalarial, antiprotozoal, antituberculosis, and antiviral activities; affecting the immune and nervous system, and other miscellaneous mechanisms of action. Comparative Biochemistry and Physiology Part C: Toxicology & Pharmacology. 2011;**153**(2):191-222

[52] Opoku G, Qiu X, et al. Effect of oversulfation on the chemical and biological properties of kappa carrageenan. Carbohydrate Polymers. 2006;**65**(2):134-138

[53] Chiu YH, Chan YL, et al. Prevention of human enterovirus 71 infection by kappa carrageenan. Antiviral Research. 2012;**95**(2):128-134

[54] Wang W, Zhang P, et al. Preparation and anti-influenza A virus activity of Î°-carrageenan oligosaccharide and its sulphated derivatives. Food Chemistry. 2012;**133**(3):880-888

[55] Li YX, Wijesekara I, et al. Phlorotannins as bioactive agents from brown algae. Process Biochemistry. 2011;**46**(12):2219-2224

[56] Miyashita K, Hosokawa M. 12 beneficial health effects of seaweed carotenoid, fucoxanthin. In: Marine Nutraceuticals and Functional Foods. New York: CRC; 2007. p. 297

[57] Hosakawa M, Bhaskar N, et al. Fucoxanthin as a bioactive and nutritionally beneficial marine carotenoid: A review. Carotenoid Science. 2006;**10**:15-28

[58] Ngo DH, Wijesekara I, et al. Marine food-derived functional ingredients as potential antioxidants in the food industry: An overview. Food Research International. 2011;**44**(2):523-529

[59] Cornish ML, Garbary DJ. Antioxidants from macroalgae: Potential applications in human health and nutrition. Algae. 2010;**25**(4):155-171

[60] Mendis E, Rajapakse N, et al. Investigation of jumbo squid (*Dosidicus gigas*) skin gelatin peptides for their in vitro antioxidant effects. Life Sciences. 2005;**77**(17):2166-2178

[61] Mendis E, Rajapakse N, et al. Antioxidant properties of a radical-scavenging peptide purified from enzymatically prepared fish skin gelatin hydrolysate. Journal of Agricultural and Food Chemistry. 2005;**53**(3):581-587

[62] Cho M, Lee HS, et al. Antioxidant properties of extract and fractions from *Enteromorpha prolifera*, a type of green seaweed. Food Chemistry. 2011;**127**(3):999-1006

[63] Raza A, Butt MS, Suleria HAR. Jamun (*Syzygium cumini*) seed and fruit extract attenuate hyperglycemia in diabetic rats. Asian Pacific Journal of Tropical Biomedicine. 2017;**7**(8):750-754

[64] Jun SY, Park PJ, et al. Purification and characterization of an antioxidative peptide from enzymatic hydrolysate of yellowfin sole (*Limanda aspera*) frame protein. European Food Research and Technology. 2004;**219**(1):20-26

[65] Sachindra NM, Sato E, et al. Radical scavenging and singlet oxygen quenching activity of marine carotenoid fucoxanthin and its metabolites. Journal of Agricultural and Food Chemistry. 2007;**55**(21):8516-8522

[66] Heo SJ, Ko SC, et al. Cytoprotective effect of fucoxanthin isolated from brown algae *Sargassum siliquastrum* against H_2O_2-induced cell damage. European Food Research and Technology. 2008;**228**(1):145-151

[67] Jung WK, Je JY, et al. A novel anticoagulant protein from *Scapharca broughtonii*. BMB Reports. 2002;**35**(2):199-205

[68] Jo HY, Jung WK, et al. Purification and characterization of a novel anticoagulant peptide from marine echiuroid worm, *Urechis unicinctus*. Process Biochemistry. 2008;**43**(2):179-184

[69] Kim SK, Pangestuti R. Potential role of marine algae on female health, beauty, and longevity. Advances in Food and Nutrition Research. 2011;**64**:41-55

[70] Mestechkina N, Shcherbukhin V. Sulfated polysaccharides and their anticoagulant activity: A review. Applied Biochemistry and Microbiology. 2010;**46**(3):267-273

[71] Maeda M, Uehara T, et al. Heparinoid-active sulphated polysaccharides from *Monostroma nitidum* and their distribution in the chlorophyta. Phytochemistry. 1991;**30**(11):3611-3614

[72] Silva F, Dore C, et al. Anticoagulant activity, paw edema and pleurisy induced carrageenan: Action of major types of commercial carrageenans. Carbohydrate Polymers. 2010;**79**(1):26-33

[73] Shanmugam M, Mody K. Heparinoid-active sulphated polysaccharides from marine algae as potential blood anticoagulant agents. Current Science. 2000;**79**(12):1672-1683

[74] Yuan YV, Walsh NA. Antioxidant and antiproliferative activities of extracts from a variety of edible seaweeds. Food and Chemical Toxicology. 2006;**44**(7):1144-1150

[75] Sithranga Boopathy N, Kathiresan K. Anticancer drugs from marine flora: An overview. Journal of Oncology. 2010;1-18. http://dx.doi.org/10.1155/2010/214186

[76] Cho EJ, Rhee SH, et al. Antimutagenic and cancer cell growth inhibitory effects of seaweeds. Preventive Nutrition and Food Science. 1997;**2**(4):348-353

[77] Yubin J, Guangmei Z. Pharmacological Action and Application of Available Antitumor Composition of Traditional Chinese Medicine. Heilongjiang, China: Heilongjiang Science and Technology Press; 1998

[78] Haroun F, Lindenmeyer F, et al. In vitro effect of fucans on MDA-MB231 tumor cell adhesion and invasion. Anticancer Research. 2002;**22**:214-221

[79] Yamamoto I, Maruyama H, et al. The effect of dietary or intraperitoneally injected seaweed preparations on the growth of sarcoma-180 cells subcutaneously implanted into mice. Cancer Letters. 1986;**30**(2):125-131

[80] Okai Y, Higashi-Okai K, et al. Identification of antimutagenic substances in an extract of edible red alga, *Porphyra tenera* (Asadusa-nori). Cancer Letters. 1996;**100**(1-2):235-240

[81] Givens DI, Gibbs RA. Current intakes of EPA and DHA in European populations and the potential of animal-derived foods to increase them: Symposium on how can the n-3 content of the diet be improved? Proceedings of the Nutrition Society. 2008;**67**(3):273-280

[82] Jiao G, Yu G, et al. Chemical structures and bioactivities of sulfated polysaccharides from marine algae. Marine Drugs. 2011;**9**(2):196-223

[83] Launay S, Hermine O, et al. Vital functions for lethal caspases. Oncogene. 2005;**24**(33):5137

[84] Fan TJ, Han LH, et al. Caspase family proteases and apoptosis. Acta Biochimica et Biophysica Sinica. 2005;**37**(11):719-727

[85] Hart AN, Zaske LA, et al. Natural killer cell activation and modulation of chemokine receptor profile in vitro by an extract from the cyanophyta *Aphanizomenon flos-aquae*. Journal of Medicinal Food. 2007;**10**(3):435-441

[86] Hiqashi-Okaj K, Otani S, et al. Potent suppressive effect of a Japanese edible seaweed, *Enteromorpha prolifera* (Sujiao-nori) on initiation and promotion phases of chemically induced mouse skin tumorigenesis. Cancer Letters. 1999;**140**(1):21-25

[87] Burja AM, Banaigs B, et al. Marine cyanobacteriaaea prolific source of natural products. Tetrahedron. 2001;**57**(46):9347-9377

[88] Vo TS, Kim SK. Potential anti-HIV agents from marine resources: An overview. Marine Drugs. 2010;**8**(12):2871-2892

[89] Adhikari U, Mateu CG. Structure and antiviral activity of sulfated fucans from *Stoechospermum marginatum*. Phytochemistry. 2006;**67**(22):2474-2482

[90] Queiroz K, Medeiros V, et al. Inhibition of reverse transcriptase activity of HIV by polysaccharides of brown algae. Biomedicine & Pharmacotherapy. 2008;**62**(5):303-307

[91] Lehtinen M, Dillner J. Preventive Human Papillomavirus Vaccination. The Medical Society for the Study of Venereal Disease; London; 2002

[92] Campo VL, Kawano DFB, et al. Carrageenans: Biological properties, chemical modifications and structural analysis: A review. Carbohydrate Polymers. 2009;**77**(2):167-180

[93] Buck CB, Thompson CD, et al. Carrageenan is a potent inhibitor of papillomavirus infection. PLoS Pathogens. 2006;**2**(7):e69

[94] Scicchitano P, Cameli M, et al. Nutraceuticals and dyslipidaemia: Beyond the common therapeutics. Journal of Functional Foods. 2014, 2014;**6**:11-32

[95] Komprda TA. Eicosapentaenoic and docosahexaenoic acids as inflammation-modulating and lipid homeostasis influencing nutraceuticals: A review. Journal of Functional Foods. 2012;**4**(1):25-38

[96] Zedler S, Faist E. The impact of endogenous triggers on trauma-associated inflammation. Current Opinion in Critical Care. 2006;**12**(6):595-601

[97] Nguyen MHT, Jung WK, et al. Marine algae possess therapeutic potential for Ca-mineralization via osteoblastic differentiation. Advances in Food and Nutrition Research. 2011;**64**:429-441

[98] Cui Y, Luo D, et al. Fucoidan: Advances in the study of its anti-inflammatory and anti-oxidative effects. Yao Xue Xue Bao [Acta Pharmaceutica Sinica]. 2008;**43**(12):1186-1189

[99] Medeiros V, Queiroz K, et al. Sulfated galactofucan from *Lobophora variegata*: Anticoagulant and anti-inflammatory properties. Biochemistry (Moscow). 2008;**73**(9): 1018-1024

[100] Ngatu NR, Okajima MK, et al. Anti-inflammatory effects of sacran, a novel polysaccharide from *Aphanothece sacrum*, on 2,4,6-trinitrochlorobenzene-induced allergic dermatitis in vivo. Annals of Allergy, Asthma & Immunology. 2012;**108**(2):117-122. e2

[101] Huang R, Zhou X, et al. Diketopiperazines from marine organisms. Chemistry & Biodiversity. 2010;**7**(12):2809-2829

[102] Heo SJ, Yoon WJ, et al. Evaluation of anti-inflammatory effect of fucoxanthin isolated from brown algae in lipopolysaccharide-stimulated RAW 264.7 macrophages. Food and Chemical Toxicology. 2010;**48**(8):2045-2051

[103] Gaven KMC, Percot A, et al. Alkaloids in marine algae. Marine Drugs. 2010;**8**(2):269-284

Chapter 3

Therapeutic Potential of Seaweed Polysaccharides for Diabetes Mellitus

Amir Husni

Additional information is available at the end of the chapter

http://dx.doi.org/10.5772/intechopen.76570

Abstract

Seaweed has attracted a great deal of interest as excellent sources of nutrients. Seaweeds contain polysaccharides, proteins, amino acids, lipids, peptides, minerals, and some vitamins. Polyphenols of seaweed was used as cosmetics and pharmacological as antioxidants, protection from radiation, anti-inflammatory, hypoallergenic, antibacterial, and antidiabetic. Besides that seaweed also has a high content of antioxidant that can be used to ward off free radicals that increase due to the condition of hyperglycemia in a patient with diabetes mellitus. Hence, a great deal of attention has been directed at isolation and characterization of seaweed polysaccharides because of their numerous health benefits, especially for diabetes mellitus. This paper is expected to provide information on the effect of alginate from two seaweeds on blood glucose and lipid profiles of diabetic rats.

Keywords: *Sargassum crassifolium*, *Turbinaria ornata*, diabetes mellitus, seaweed, alginate

1. Introduction

Diabetes mellitus (DM) is a disease caused by hyperglycemia due to a relative or absolute insulin insufficiency. Chronic hyperglycemia can cause complications such as neuropathy, retinopathy, nephropathy, and cardiovascular disease [1]. Hyperglycemia can also cause impaired balance metabolism of carbohydrates, fats, and proteins [2]. International Diabetes Federation (IDF) estimates that in 2013 there were 382×10^6 people with diabetes and 316×10^6 people suffer from impaired glucose tolerance and increased risk of diabetes. These results are expected to increase to 471×10^6 in 2035 and predicted less than 25 years; there would be 592×10^6 people have diabetes without quick and precise prevention [3].

Seaweeds are the most abundant resources in the ocean. Seaweeds contain polysaccharides, proteins, amino acids, lipids, peptides, minerals, and some vitamins. Polyphenols of seaweed was used as cosmetics and pharmacological as antioxidants, protection from radiation, anti-biotics, anti-inflammatory, hypoallergenic, antibacterial, and antidiabetic [4]. Polyphenol extracts from seaweed, for example, *Alaria, Ascophyllum, Padina,* and *Palmaria,* are able to inhibit the activity of α-amylase and α-glucosidase that can lower blood glucose levels [5, 6]. On the other hand, seaweed also has a high content of antioxidants that can have beneficial value for diabetes mellitus patient [7]. Research on the use of Na alginate from *Turbinaria ornata* and *Sargassum crassifolium* on in vivo studies in diabetic rats was limited. This paper is expected to provide information about the effect of Na alginate from *T. ornata* and *S. crassifo-lium* on the blood glucose and lipid profiles of drug-induced diabetic rats.

2. Extraction of polysaccharides from marine algae

Na alginate from *T. ornata* and *S. crassifolium* was extracted as explained by Husni et al. [8, 9]. Dried samples were weighted and were soaked in distilled water with the addition of 0.1 N HCl to pH 4 for about 24 h 1:15 (w/v). The seaweed was washed with distilled water until pH 7. The filtrate was added with 0.5 N Na_2CO_3 (pH 11) 1:10 (w/v) and then heated at 60°C for 2 h. The viscous mixture was added with distilled water 1:10 (w/v) and separated from its residue by centrifuge (3500 rpm, 5 min, 4°C) (1 rpm = 1/60 Hz). The Na alginate extract was added with 5 N H_2O_2 1:4 (v/v), stirred for 30 min before left for two h. The mixture was added with 0.5 M $CaCl_2$ and stirred for 30 min followed by adding 0.5 N HCl until pH 2. The mixture was stirred and left for 30 min at room temperature. Insoluble material (alginic acid) was separated from the supernatant by centrifuge. Alginic acid was weighed, was added with distilled water and 0.5 N Na_2CO_3 2:2:3 (w/v/v), and was stirred for one h at room temperature to obtain a solid form of Na alginate. Na alginate was precipitated with EtOH slowly 1:1 (v/v) and stirred for 30 min, after being centrifuged, followed by drying at 60°C, and the yield of alginate was determined.

3. Alginate characterizations (structural and physical properties)

FTIR spectroscopy was used to identify the polysaccharide structures. A pellet of sodium alginate was prepared with KBr. FTIR spectrum was recorded on Shimadzu-FTIR Prestige 21 with a resolution of 4 cm^{-1} in the 4000–400 cm^{-1} region, with a scan speed of 0.20 cm s^{-1}. The FTIR spectrum of sodium Na alginate of *T. ornata* showed similar bands to that Na alginate standard in 3500–1300 cm^{-1} region, while the fingerprint region has two bands at 948.98 cm^{-1} and 871.82 cm^{-1} (**Figure 1**). Sodium alginate of *T. ornata* showed eight charac-teristic bands which also could be found in sodium alginate standard (**Table 1**). According to literature, the band at 3400 cm^{-1} assigned to the hydrogen bonded O-H stretching vibra-tions and the weak signal at 2931.80 cm^{-1} due to C-H stretching vibrations [10] and the asymmetric stretching of carboxylate O-C-O vibration at 1627.92 cm^{-1} [10, 11]. The band at

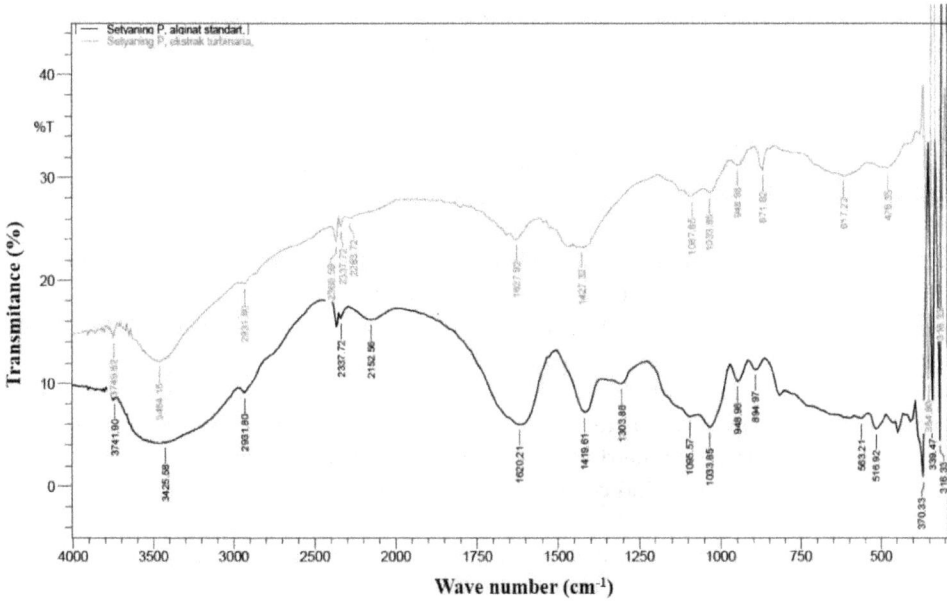

Figure 1. Infrared spectra of Na alginate standard (red) and Na alginate of *T. ornata* (black).

Wave number (cm⁻¹)		Assignments
Na alginate standard [8]	Na alginate of *T. ornata* [8]	
3425.58	3464.15	vO–H a
2931.80	2931.80	vC–H a
1620.21	1627.92	v asym COO- b
1419.61	1427.32	δ C–O–H, vsym COO– (carboxylate ion)[1,3]
1095.57	1087.85	v C-O, v C-C (pyranose ring)[1,2,3]
1033.85	1033.85	v C-O[1]
948.98	948.98	v C-O (uronic acid residues)[1,3]
894.97	871.82	δ C1-H (β-mannuronic residues)[1,3]

[1]Source: [10].
[2][11]
[3][12]

Table 1. FTIR spectrum of Na alginate from *T. ornata* and standard.

1427.32 cm⁻¹ is assigned to C-O-H deformation vibration with the contribution of O-C-O symmetric stretching vibration of carboxylate group [10, 12]. The band 1087.85 cm⁻¹ might be assigned to C-O and C-C stretching vibrations of pyranose ring [10–12], and the band at 1033.85 cm⁻¹ might also be due to C-O stretching vibrations [10]. The anomeric region of the fingerprint (950–750 cm⁻¹) showed two characteristic absorption bands. The band at 948.98 cm⁻¹ was assigned to the C-O stretching vibration of uronic acid residues, and the one at 871.82 cm⁻¹ was assigned to the C1-H deformation vibration of β-mannuronic acid residues [10, 12].

The peak infrared spectrum of standard alginate and *S. crassifolium* can be seen in **Table 2**. Based on the FTIR test conducted on alginate extract of *S. crassifolium* and alginate standard (**Figure 2**) on the first band of alginate extract spectra, the vibration frequency of 779.24 cm⁻¹ shows the residue of guluronic acid. The second band, standard alginate spectra and alginate extracts, contained the same vibration frequency at 948.98 cm⁻¹ showing the suspected vibration of C-O stretching as uronic acid. The third band detected vibrations from C-O stretching, wavelengths 1033.85 cm⁻¹ at standard alginate, and 1026.13 cm⁻¹ on alginate extract. The fourth band in the standard alginate detected vibrations at wavelengths of 1095.57 and 1087.85 cm⁻¹ in the extra alginate indicating the presence of OCO rings. Symmetrical and asymmetrical C-O vibrations were detected in the standard alginate of the fifth and sixth bands indicating the presence of carboxylic groups at 1303.88 and 1419.61 cm⁻¹ wavelengths, but this vibration was not detected in the alginate extract. The seventh band contained a vibration of 2931.8 cm⁻¹ in standard alginates and alginate extracts indicating the presence of C-H stretching. O-H is stretching vibration indicating the presence of H atomic bonds detected in both alginates, 3425.58 cm⁻¹ in standard alginate and 3471.87 cm⁻¹ in alginate extract.

Wavelength (cm⁻¹)			Type of vibration
Na alginate standard [9]	Na alginate of *S. crassifolium* [9]	Reference	
—	779.24	778.20[1]	Guluronic acid residues
948.98	948.98	950–810[2]	C-O stretching
1033.85	1026.13	1023.40[1]	C-O stretching
1095.57	1087.85	1100–1050[3]	OCO ring (shoulder)
1303.88	—	1320–1210[2]	C-O stretching
1419.61	—	1460–1400[2]	C-O asymmetric stretching
2931.80	2931.80	~2925[2]	C-H stretching
3425.58	3471.87	3600–3200[3]	O-H stretching

[1]Source: [13].
[2][14].
[3][15].

Table 2. FTIR spectrum of Na alginate from *S. crassifolium* and standard.

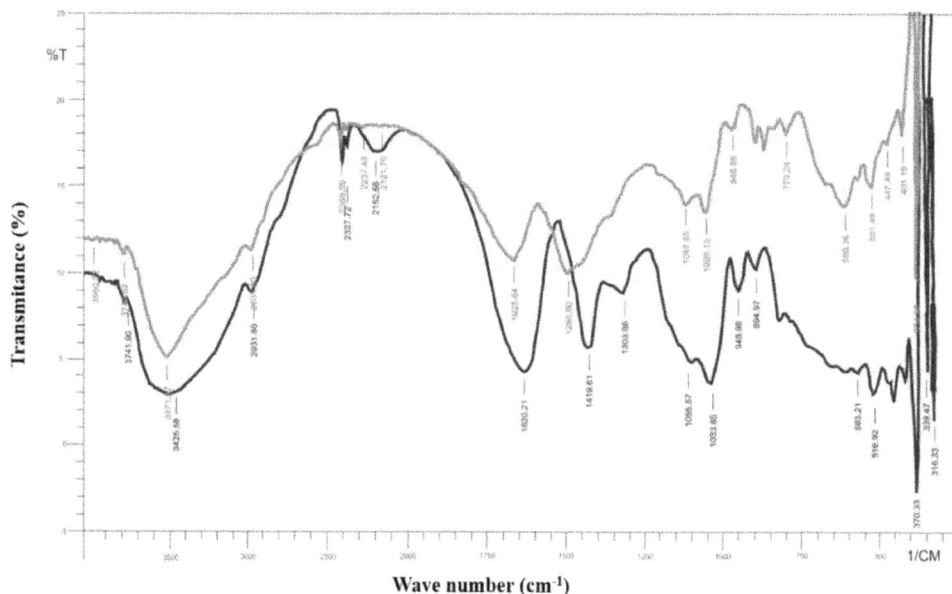

Figure 2. Infrared spectra of Na alginate standard (red) and Na alginate of *S. crassifolium* (black).

4. Biological activity of polysaccharides from marine algae

4.1. Effect of Na alginate of *T. ornata* on body weight of rats

Alloxan-induced diabetic rats did not show a significant decrease in body weight after the injection of alloxan. Five groups of diabetic rats had decreased in body weight on 15 days treatment, and there were significant differences between the groups of rats. There was no significant difference between diabetic control (negative control) compared to positive control, and the positive control was not significantly different compared to alloxan diabetic rats treated with Na alginate 200 mg/kg. Alloxan-induced diabetic rats treated with Na alginate(s) (200, 400, 600 mg/kg) did not show significant difference between each other. Administration of Na alginate(s) (400, 600 mg/kg) showed a significant difference compared to negative control. The body weight of alloxan-induced diabetic rats treated with Na alginate 600 mg/kg was not significantly different compared to normal control.

The lowering of rats' body weight treated with alginate from *T. ornata* showed lower than a study conducted by Wikanta et al. [16] using κ-carrageenan and ι-carrageenan. In those researches, κ-carrageenan increased the weight by 34.1 g, and ι-carrageenan increased the weight by 30.1 g from the body weight on alloxan-induced diabetic rats after 15 days of treatment. The significant reduction in total body weight could be attributed to the loss of fat from adipose tissue and catabolism of amino acids in the muscle tissue [17].

4.2. Effect of Na alginate of *S. crassifolium* on body weight of rats

Diabetic mice showed weight loss in all treatment groups except the normal control group. Normal control group gained weight of 24.1 g. The negative control group had a very signifi-cant weight loss of 51.6 g. The positive control group had a weight loss of 47.2 g. The treat-ment group of extract 200 mg/kg had a weight loss of 58.8 g. The treatment group of 400 mg/ kg extract had a weight loss of 45.3 g. Meanwhile, the treatment group giving 600 mg/kg extract experienced a decrease in body weight by 43.1 g. Streptozotocin (STZ)-induced dia-betic rats are one of the animal models of type 1 diabetes mellitus. It is well known for its selective pancreatic islet beta-cell cytotoxicity and has been extensively used to induce type 1 diabetes in an experimental rat model. Glibenclamide is often used as a standard antidiabetic drug in STZ-induced diabetes to compare the efficacy of a variety of hypoglycemic drugs [18].

Throughout the experiments, all the rats were monitored daily and/or weekly for the symp-toms of type 1 diabetes mellitus, including polydipsia, polyuria, polyphagia, hyperglycemia, and muscle wasting leading to weight loss and insulin deficiency. **Figure 1** shows the observa-tions of body weight of treated rats during the whole period of experiments. The body weight was continuously increased in the normal group and decreased in all diabetes groups. A severe loss of body weight characterizes STZ-induced diabetes. Due to absolute or relative deficiency of insulin and decrease of the production of ATP, protein synthesis decreases in all tissues.

5. Effect of Na alginate on blood glucose

Alloxan is a urea derivative which causes selective necrosis of the pancreatic islet β-cells [19]. Alloxan and its reduction product dialuric acid establish a redox cycle with the formation of superoxide radicals [20]. Preprandial blood glucose levels were determined as fasting blood glucose. Fasting is defined as no calorie intake for at least 8 h [1]. Diabetes is diagnosed when the fasting plasma glucose concentration is consistently ≥ 7 mmol/L (126 mg/dL) or when the 2 h plasma glucose concentration (after drinking a 75 g glucose load) is consistently ≥ 11.1 mmol/L (200 mg/dL) [21].

Administration of alloxan led to a significant increase of preprandial blood glucose levels in rats after 3 days. Administration of Na alginate(s) (200, 400, 400 mg/kg) significantly reduced the blood glucose level compared to diabetic control. The dose of 200 and 400 mg/ kg of Na alginate did not show a significant difference compared to normal control and positive control (**Table 3**). The result was supported by previous studies using fiber to decrease preprandial blood glucose. Nelson et al. [22] used high indigestible fiber and low indigestible fiber diet to decrease preprandial blood glucose in diabetic dogs for 8 months which resulted in high indigestible fiber significantly that reduces preprandial blood glu-cose better than low indigestible fiber. Nelson et al. [23] used similar treatment in diabetic cats for 24 weeks and showed high indigestible fiber which gave a better effect on decreas-ing preprandial blood glucose than low indigestible fiber. Chandalia et al. [12] compared the amount of fiber that was given to diabetic patients according to the American Diet

Group	Preprandial blood glucose (mg/dL)*	
	T. ornata [8]	S. crassifolium [9]
Normal control	106.06 ± 11.33[b,c]	126.30 ± 0.50[a]
Negative control	208.57 ± 70.60[a]	568.82 ± 46.40[c]
Positive control	86.29 ± 13.83[b]	316.35 ± 20.90[b]
Alginate (200 mg/kg)	108.50 ± 11.28[c]	279.45 ± 92.50[b]
Alginate (400 mg/kg)	96.55 ± 15.65[b,c]	336.63 ± 66.32[b]
Alginate (600 mg/kg)	99.03 ± 14.26[b,c]	257.66 ± 34.61[b]

*Values are means ± SD. Values followed by the different superscript symbol(s) in each column were significantly different ($P<0.05$).

Table 3. Effect of Na alginate of *T. ornata* and *S. crassifolium* on preprandial blood glucose in alloxan-induced diabetic rats.

Association (8 g digestible fiber and 16 indigestible fiber) and fiber-rich diet (25 g digestible fiber and 25 indigestible fiber) for 6 weeks. Fiber-rich diet decreased 13% preprandial blood glucose lower than ADA diet.

Normal postprandial blood glucose level is <180 mg/dL [1]. In the normal state, the postprandial blood glucose level increases less than 50 mg/dL from the preprandial blood glucose level after carbohydrate intake [24]. Alloxan-induced diabetic rats' postprandial blood glucose level surpassed 200 mg/dL after 3 days of injection. After 15 days of treatment, the result was the administration of Na alginate(s) (200, 400, 600 mg/kg) which significantly reduces postprandial blood glucose levels on rats compared to diabetic control ($P < 0.05$). However, it failed to restore the level to that of normal control group and positive control group ($P < 0.05$). The positive control group could restore the postprandial blood glucose level at the same level as a normal control group (**Table 4**).

Wolf et al. [25] used 1.5 g sodium alginate to show its effect on postprandial glucose peak and glucose uptake reduction after 3 h which resulted in line 32.80 ± 3.40 and 1429 ± 276 mg/dL. Sodium alginate had a reduction effect better than 1.2 g gum arabic and 0.3 g gum guar with postprandial glucose peak 40.40 ± 3.30 mg/dL and glucose uptake 1717 ± 433 mg/dL. A study on the effect of a meal containing alginate compared to testing a meal without alginate by Torsdottir et al. [26] showed that postprandial blood glucose levels by meal containing alginate decrease 31% lower than a meal without alginate.

Preprandial glucose levels for all treatment groups of alginate from *S. crassifolium* were classified into normal levels ranging from 69.311 to 88.029 mg/dL and no significant difference. The streptozotocin-induced treatment group experienced very high preprandial glucose levels exceeding 200 mg/dL and can be categorized as DM. The same is also shown in Moree et al. [27]. In this study, male Wistar rats induced by streptozotocin dose 60 mg/kg increased blood glucose levels <200 mg/dL. The use of 60 mg/kg of streptozotocin in mice can trigger an autoimmune process that can produce damage to the Langerhans island beta cells [28]. Also,

Group	Postprandial blood glucose (mg/dL)*	
	T. ornata [8]	S. crassifolium [9]
Normal control	133.05 ± 15.81[b]	150.416 ± 5.1[a]
Negative control	360.48 ± 40.80[a]	633.470 ± 27.8[c]
Positive control	140.75 ± 9.16[b]	333.814 ± 64.5[ab]
Alginate (200 mg/kg)	257.08 ± 34.20[c]	421.652 ± 21.4[bc]
Alginate (400 mg/kg)	238.61 ± 21.48[c]	433.333 ± 21.8[bc]
Alginate (600 mg/kg)	196.05 ± 18.22[d]	381.250 ± 11.4[ab]

*Values are means ± SD. Values followed by the different superscript symbol(s) in each column were significantly different ($P<0.05$).

Table 4. Effect of Na alginate of T. ornata and S. crassifolium on postprandial blood glucose in alloxan-induced diabetic rats.

STZ is also capable of generating reactive oxygen that has a high role in the destruction of pancreatic β-cells and eventually occurs inhibition of insulin secretion and synthesis resulting in hyperglycemia [29].

All treatment groups of extracts of S. crassifolium did not differ significantly with the positive control (**Table 3**). It can be concluded that all three doses of administration of the extract of S. crassifolium had the same effect as the positive control group in lowering blood glucose levels in mice suffering from DM. The opposite is shown in the negative control treatment group that has a prepreg glucose level that increases from day to day due to accumulated glucose buried in the blood without treatment efforts.

In general, the viscosity of dietary fiber can reduce the rise in blood glucose levels and reduce food intake by slowing the empty stomach and slowing the absorption of nutrients in the small intestine. Based on these two mechanisms, it is still not clear what mechanisms apply to sodium alginate, perhaps one or both [30]. Different doses of alginate will affect the viscosity of the given test preparation. So, it will lead to differences in the viscosity of the fluid in the gastrointestinal tract and ultimately result in differences in the rate of glucose absorption from the gastrointestinal tract into the blood vessels [31].

6. Total cholesterol

Diabetes is associated with major abnormalities in fatty acid metabolism. The resulting disturbance results in an abnormal lipoprotein cascade from the large chylomicron through to the small HDL particle [31, 32]. Total cholesterol in the serum of negative control was not significantly different compared to positive control, Na alginate 200 and 400 mg/kg treatment, and normal control. Na alginate 600 mg/kg of T. ornata was a significant difference compared to negative control (P < 0.05). The alginate dose of 200 and 600 mg/kg of T. ornata did not show the difference (P > 0.05) (**Table 5**) significantly.

Several previous studies supported the result. Suzuki et al. [33] evaluated the effect of algi-nate-rich guluronic and mannuronic on cholesterol levels in rats fed with diets containing both alginates and cholesterol which resulted from reductions in liver cholesterol in rats fed with each alginate and significantly low cholesterol accumulation in mannuronic acid-rich alginate. Ren et al. [34] screened 26 species of seaweeds and six polysaccharides from algae to study their effect on lipid in rats fed with basal diet for 28 days of treatment. The six polysaccharides were sulfated glucuronoxylomannan (0.5%), fucoidin (1%), sodium alginate (1%), funorin (2.5%), porphyrin (2.5%), and agar (2.5%). Reduction effect of each polysac-charide was 64, 65, 68, 77, 88, and 95%, respectively, compared to control group. At the end of the study, the polysaccharides could restore the cholesterol level to the same level as the control group.

Total cholesterol levels of the normal control group, positive control, and alginate 600 mg/kg of S. crassifolium had a significant difference compared to the negative control group (**Table 5**). The three treatment groups had lower cholesterol levels than the negative control group. An extract at a dose of 600 mg/kg of S. crassifolium can lower total cholesterol levels as well as positive controls (glibenclamide). The opposite is shown by the treatment group giving the extract dose of 200 and 400 mg/kg of S. crassifolium. Both the doses are less effective in lower-ing total cholesterol levels in mice suffering from diabetes compared to glibenclamide and alginate 600 mg/kg of S. crassifolium.

Wikanta et al. [35] reported that sodium alginate could lower total cholesterol in mice with hypercholesterolemia. Administration of sodium alginate with a viscosity of 450 cps signif-icantly reduced total cholesterol levels compared to sodium alginate with lower viscosity. Because, sodium alginate is a water-soluble fiber compound, forming a viscous solution. The stomach fluid cannot digest this compound in the gastrointestinal tract. When dissolved in water, the sodium alginate fibers form a mesh-like grid that strongly binds many water mol-ecules in a well-defended solute. Its properties as emulgator increasingly enhance the binding ability. A similar mechanism occurs against lipid molecules in bile acids in the gastrointesti-nal tract. The binding or bonding of lipids by the alginate makes lipid and cholesterol unable

Group	Total cholesterol (mg/dL)*	
	T. ornata [8]	S. crassifolium [9]
Normal control	70.40 ± 7.12^b	41.55 ± 0.20^a
Negative control	67.75 ± 16.02^b	68.41 ± 12.50^b
Positive control	72.40 ± 15.24^b	45.79 ± 9.80^a
Alginate (200 mg/kg)	$55.80 \pm 3.42^{a,b}$	49.05 ± 20.00^{ab}
Alginate (400 mg/kg)	65.60 ± 14.47^b	54.46 ± 11.00^{ab}
Alginate (600 mg/kg)	47.80 ± 5.40^a	34.20 ± 7.50^a

*Values are means ± SD. Values followed by the different superscript symbol(s) in each column were significantly different ($P<0.05$).

Table 5. Effect of Na alginate of T. ornata and S. crassifolium on the total cholesterol in alloxan-induced diabetic rats.

to absorb the body through the small intestine so that it eventually comes out with the stool. Suzuki et al. [33] also reported that alginate with various mannuronic acid and guluronic acid compositions can decrease total blood cholesterol levels.

7. HDL-c

Administration of Na alginate to alloxan-induced diabetic rats for 200 mg/kg alginate of *T. ornata* did not show significant differences compared to negative control and positive control (P > 0.05) (**Table 6**). The alginate of *T. ornata* at a dose of 200 and 400 mg/kg was not significantly different between each other. All of the various doses of alginate were significantly different compared to normal control (P < 0.05). HDL-c management on type 2 diabetes is targeting for >40 mg/dL (>50 mg/dL on female) [1]. HDL particles seem to have antioxidant properties, inhibiting the oxidation of LDL cholesterol and the expression of cellular adhesion molecules and monocyte recruitment. The HDL may also reduce the risk of thrombosis by inhibiting platelet activation and aggregation [33]. Ren et al. [34] reported that three algal species showed the ability to increase HDL-c levels in blood serum of rats. Fucoidan could increase HDL-c levels up to 47% compared to the control group. Five other polysaccharides, sulfated glucuronoxylorhamman, sodium alginate, funoran, porphyran, and agar, found increased HDL-c by 31.97, 28.93, 9.14, 3.55, and 26.90%, respectively.

According to Rohman [36] HDL is a protective lipoprotein, in addition to functioning to bring fat to the liver; HDL proved to inhibit the oxidation of LDL and adhesion molecules. HDL-c levels throughout the treatment group did not have a significant difference. The same is also shown in the study of Suzuki et al. [33] that there was no statistically significant difference in HDL-c levels in mice suffering from hypercholesterolemia treated with sodium alginate in comparison with different glucuronic acid and mannuronic acids.

Group	HDL-c (mg/dL)*	
	T. ornata [8]	*S. crassifolium* [9]
Normal control	108.00 ± 6.59[c]	70.549 ± 1.50[a]
Negative control	59.75 ± 9.39[a]	75.549 ± 11.10[a]
Positive control	58.00 ± 7.78[a]	96.843 ± 14.10[a]
Alginate (200 mg/kg)	61.80 ± 5.57[a]	97.617 ± 11.50[a]
Alginate (400 mg/kg)	74.80 ± 10.08[b]	84.03 ± 28.20[a]
Alginate (600 mg/kg)	78.60 ± 10.60[b]	75.98 ± 17.70[a]

*Values are means ± SD. Values followed by the different superscript symbol(s) in each column were significantly different (*P*<0.05).

Table 6. Effect of Na alginate of *T. ornata* and *S. crassifolium* on HDL-c in alloxan-induced diabetic rats.

8. LDL-c

LDL-c after administration of alginate(s) from *T. ornata* (200, 400, 600 mg/kg) was not significantly different between each other. Alginate of *T. ornata* 600 mg/kg showed a significant difference compared to negative control, positive control, and normal control group (**Table 7**). Ren et al. [34] studied the effect of polysaccharide extracts from algae on LDL-c in blood serums of rats given with basal diet for 28 days. The six polysaccharides used in the study decreased LDL-c levels in blood serum. Sodium alginate (1%) decreased 34.04% of LDL-c. Five other polysaccharides, sulfated glucuronoxylorhamman, sodium alginate, funoran, porphyran, and agar, decreased the LDL-c in line with 36.42, 37.66, 24.33, 36, and 14%, respectively, compared to normal control. LDL is not usually increased in diabetes. In part, this may represent a balance of factors that affect LDL production and catabolism. A necessary step in LDL production is hydrolysis of its precursor VLDL by LpL. A reduction can happen in this step because LpL deficiency or excess surface apoproteins (C1, C3, or possibly E) decreases LDL synthesis. Conversely, increases in this lipolytic step that accompany weight loss, fibric acid drug therapy, and treatment of diabetes may increase LDL levels. In diabetes, a reduction in LDL production may be counterbalanced by decreases in LDL receptors and/or the affinity of LDL for those receptors [37].

Administration of sodium alginate from *S. crassifolium* most effective in lowering LDL-c levels near the control group was 600 mg/kg followed by 200 mg/kg and 400 mg/kg. The negative control group had a very significant difference with all the other treatment groups (**Table 7**). The negative control group had higher LDL-c levels when compared to the other treatment groups. Meanwhile, the positive control group had lower LDL-c levels than the other

Group	LDL-c (mg/dL)*	
	T. ornata [8]	*S. crassifolium* [9]
Normal control	58.80 ± 7.19[a]	34.07 ± 0.90[a]
Negative control	60.75 ± 16.52[a]	55.34 ± 8.30[b]
Positive control	65.00 ± 14.05[a]	27.51 ± 10.00[a]
Alginate (200 mg/kg)	49.60 ± 3.13[a,b]	31.81 ± 11.80[a]
Alginate (400 mg/kg)	55.60 ± 13.13[a,b]	33.91 ± 5.30[a]
Alginate (600 mg/kg)	41.00 ± 5.83[b]	28.78 ± 5.30[a]

*Values are means ± SD. Values followed by the different superscript symbol(s) in each column were significantly different ($P<0.05$).

Table 7. Effect of Na alginate of *T. ornata* and *S. crassifolium* on LDL-c in alloxan-induced diabetic rats.

treatment groups. Levels of LDL-c in this study are still within normal limits, i.e., <130 mg/dL, and Rachmat and Rasyid [38] reported that mice were given 50 and 250 g of alginates of *S. crassifolium* which also did not affect LDL-c levels.

9. Triglyceride

Triglyceride management on type 2 diabetes is targeting for <150 mg/dL [1]. When the glucose levels excess in the blood, glucose will be converted to triglycerides in which triacylglycerol synthesis process is known as lipogenesis. Carbohydrate-rich meal can lead to increase the process of lipogenesis in the liver and adipose tissue. However, the occurrence of insulin resistance inhibits lipogenesis process making glucose and free fatty acid levels in blood plasma increased. In the liver, triglyceride accumulation can cause malfunctioning of the liver (fatty liver) or liver cirrhosis in the long term [39]. Triglyceride of alloxan-induced diabetic rats did not show a significant difference between the groups of treatment using alginate of *T. ornata*. The triglyceride levels remained at normal levels through the given time of the study (**Table 8**).

Paxman et al. [40] reported that a drink containing alginate in the obese patient had no effect on tryglyceride level. Triglyceride levels did not show a significant difference between alginate treatment group and control group. Ren et al. [34] used six polysaccharides from algal species as a treatment for rats given with basal diet for 28 days. All of the polysaccharides used in this research could reduce triglyceride levels as good as their ability reducing LDL-c in blood serum. Funoran and sulfated glucuronoxylorhamman reduced triglyceride levels between 46 and 64% compared to the control group. Sodium alginate could decrease the

Group	Triglyceride (mg/dL)*	
	T. ornata [8]	*S. crassifolium* [9]
Normal control	75.80 ± 10.33[a]	28.73 ± 12.20[a]
Negative control	77.75 ± 20.90[a]	77.73 ± 14.10[b]
Positive control	80.40 ± 13.14[a]	24.31 ± 9.60[a]
Alginate (200 mg/kg)	63.40 ± 25.41[a]	24.12 ± 17.70[a]
Alginate (400 mg/kg)	60.80 ± 13.80[a]	31.73 ± 2.90[a]
Alginate (600 mg/kg)	54.80 ± 10.91[a]	37.67 ± 8.50[a]

*Values are means ± SD. Values followed by the different superscript symbol(s) in each column were significantly different ($P<0.05$).

Table 8. Effect of Na alginate of *T. ornata* and *S. crassifolium* on the triglyceride in alloxan-induced diabetic rats.

triglyceride level to 29% compared to the control group. Fucoidan can reduce the triglyceride levels to 12–20% [34].

The levels of triglycerides during the experiment using alginate of *S. crassifolium* were decreased. The negative control treatment group had a significant difference when compared to all treatment groups (**Table 8**). This suggests that the three doses of alginate from *S. crassifolium* can lower triglyceride levels equally well with the positive control group that is close to the triglyceride levels of the normal control group.

All groups treated with DM except for the normal control group showed elevated triglyceride levels. Levels of triglycerides increased up to 574.867 mg/dL. The condition of hypertriglyceridemia can be diagnosed if the triglyceride level >150 mg/dL [41]. According to Pujar et al. [42], this can be due to direct damage from the pancreatic tissue by high free fatty acids. The concentration of high free fatty acid will decrease the pH and may activate trypsinogen. Also, high triglyceride levels can also be caused by the destruction of chylomicron which is a triglyceride carrier. This changes the acinar function and opens the pancreatic tissue to triglycerides.

10. Necrosis of pancreas

Necrosis is defined as the type of cell death caused by changing the morphology of the nucleus, including chromatin condensation and fragmentation, minor changes in cytoplasmic organelles, and overall causes of cell shrinkage (apoptosis) and autophagic accumulation of two vacuole membranes in the cytoplasm [43]. In type I diabetes mellitus, patients found changes in the pancreas in the form of the reduced size of the pancreas, atrophy in the exocrine pancreas, and atrophy of the acinar cells around the degenerated Langerhans island. On the other hand, in type II diabetes mellitus, an imbalance of exocrine secretion of the pancreas and impaired control of blood glucose occur [44].

Normal controls show normal cell conditions (**Figure 3**). Negative controls show some damage to the cell. The positive control treatment group also shows the same. The treatment group of sodium alginate extract is entirely damaged in cells (necrosis). The treatment group of *S. crassifolium* dose of alginate at 200 and 400 mg/kg had more damage than the treatment group of 600 mg/kg alginate. The results of the histological analysis showed that all treatment groups experienced cell damage (necrosis) except the normal control group. According to Holemans et al. [45], streptozotocin prevents DNA synthesis in mammals and bacterial cells. In bacterial cells, it provides a special reaction with the cytosine group that causes degeneration and destruction of DNA. This biochemical reaction in mammals causes cell death. Damage to cells in the islets of Langerhans island cells caused by streptozotocin is irreversible. Similar results were also shown in a study conducted by Elias et al. [46] and Ikebukuro et al. [47].

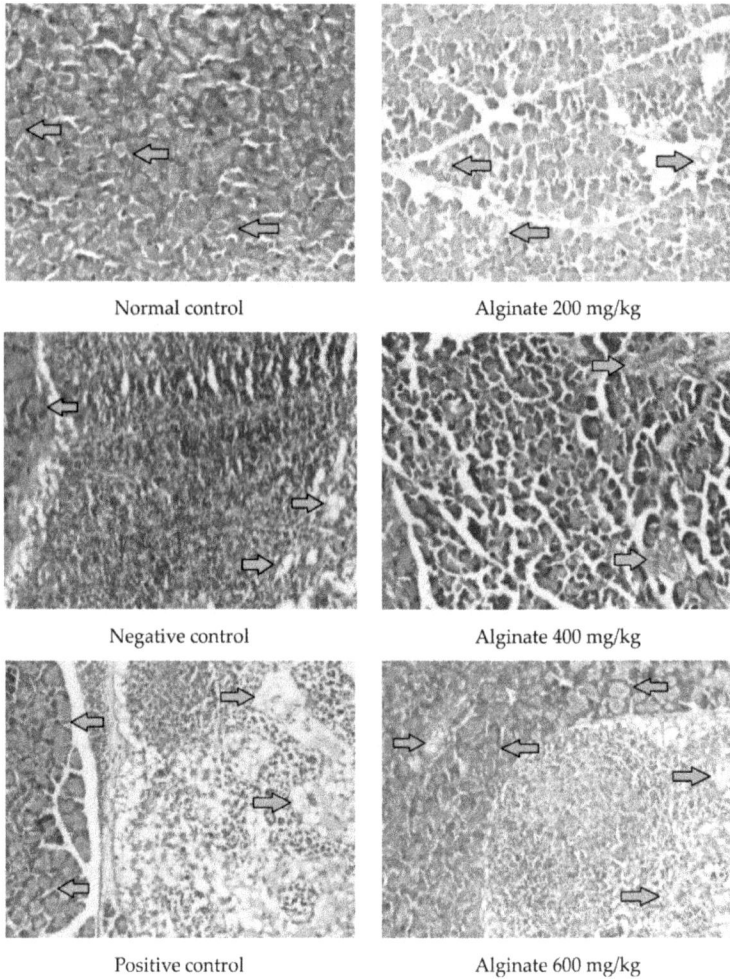

Figure 3. Histological studies of STZ diabetic rat pancreas. Normal control: pancreatic section showed the normal size of islets, and destruction was absent (Grade -). Negative control: pancreatic section showed (green arrow) occasional islets, and (orange arrow) destruction was severe (Grade ++++). Positive control (diabetic rats +5 mg glibenclamide/kg b.w.): pancreatic section showed moderate islet architecture (green arrow), and destruction (orange arrow) was moderate (Grade +++). Diabetic rats +200 mg alginate/kg b.w., and diabetic rats 400 mg alginate/kg b.w.: pancreatic section showed (green arrow) occasional islets, and (orange arrow) destruction was severe (Grade ++++). Diabetic rats +600 mg alginate/kg b.w.: pancreatic section showed (green arrow) additive improvement in the mass of islets as compared to other alginate treatments, and (orange arrow) destructions was mild (Grade ++). Grade -, normal; Grade ++++, severe destruction; Grade +++, moderate destruction; Grade ++, mild injury.

11. Conclusion

Administration of alginate from *T. ornata* in alloxan-induced diabetic rats decreased the pre-prandial and postprandial blood glucose, lowered total cholesterol, increased HDL-c, and lowered LDL-c in dependent dose manner. However, sodium alginate of *T. ornata* did not show any effect on triglyceride. This result can be valuable information to discover alternative therapy to achieve and/or maintain glycemic control and lipid profile management on diabetes patient. Nevertheless, the possibility warrants further confirmation. On the other hand, the present study shows that the alginate *S. crassifolium* has potential antidiabetic action in STZ-induced diabetic rats and the effect was found to be more similar to the reference drug glibenclamide.

Acknowledgements

Research Grants Flagship Universitas Gadjah Mada supported this research through DIPA UGM 2014 number LPPM-UGM/478/LIT/2014.

Conflict of interest

The authors declare no conflict of interest.

Author details

Amir Husni

Address all correspondence to: a-husni@ugm.ac.id

Department of Fisheries, Faculty of Agriculture Universitas Gadjah Mada, Yogyakarta, Indonesia

References

[1] American Diet Association (ADA). Diagnosis and classification of diabetes mellitus. Diabetes Care. 2012;**35**:64-71. DOI: 10.2337/diacare.27.2007.S5

[2] Boden G, Laakso M. Lipids and glucose in type 2 diabetes. Diabetes Care. 2004;**27**:2253-2259. DOI: 10.2337/diacare.27.9.2253

[3] International Diabetes Federation (IDF). IDF Diabetes Atlas. 6th ed (online version). 2013. Available from: https://idf.org/e-library/epidemiology-research/diabetes-atlas/19-atlas-6th-edition.html [Accessed: February 23, 2018]

[4] Gamal E. Biological importance of marine algae. Saudy Pharmacy Journal. 2010;**18**:1-25. DOI: 10.1016/j.jsps.2009.12.001

[5] Nwosu F, Morris J, Lund VA, Stewart D, Ross HA, McDougall GJ. Anti-proliferative and potential anti-diabetic effects of phenolic-rich extracts from edible marine algae. Food Chemistry. 2011;**126**:1006-1012. DOI: 10.1016/j.foodchem.2010.11.111

[6] Husni A, Wijayanti R, Ustadi. Inhibitory activity of α-amylase and α-glucosidase by *Padina pavonica* extracts. Journal of Biological Sciences. 2014;**14**:515-520. DOI: 10.3923/jbs.2014.515.520

[7] Firdaus M, Astawan M, Muchtadi D, Wresdiyati T, Waspadji S, Karyono SS. Prevention of endothelial dysfunction in streptozotocin-induced diabetic rats by *Sargassum echino-carpum* extract. Medical Journal of Indonesia. 2010;**19**:32-35. DOI: 10.13181/mji.v19i1.382

[8] Husni A, Pawestria S, Isnansetyo A. Blood glucose level and lipid profile of alloxan-induced diabetic rats treated with Na alginate from seaweed *Turbinaria ornata* (Turner) J. Agardh. Jurnal Teknologi. 2016;**78**(4-2):7-14. DOI: 10.11113/jt.v78.8145

[9] Husni A, Purwanti D, Ustadi. Blood glucose level and lipid profile of streptozotocin-induced diabetes rats treated sodium alginate from *Sargassum crassifolium*. Journal of Biological Sciences; **16**:58-64. DOI: 10.3923/jbs.2016.58.64

[10] Leal D, Matsuhiro B, Rossi M, Caruso F. FT-IR spectra of alginic acid block fractions in three species of brown seaweeds. Carbohydrate Research. 2008;**343**:308-316. DOI: 10.1016/j.carres.2007.10.016

[11] Campos-Vallette MM, Chandía NP, Clavijo E, Leal D, Matsuhiro B, Osorio-Rom'an IO, Torres S. Characterization of sodium alginate and its block fractions by surface-enhanced raman spectroscopy. Journal of Raman Spectroscopy. 2010;**41**:758-763. DOI: 10.1002/jrs.2517

[12] Chandalia M, Garg A, Lutjohanh D, von-Bergmann K, Grundy SM, Brinkley LJ. Beneficial effects of high dietary fiber intake in patients with type 2 diabetes mellitus. The New England Journal of Medicine. 2000;**324**:1392-1398. DOI: 10.1056/NEJM200005113421903

[13] Sergios KP, Kouvelos EP, Favvas EP, Sapalidis AA, Romanos GE, Katsaros FK. Metal–carboxylate interactions in metal–alginate complexes studied with FTIR spectroscopy. Carbohydrate Research. 2010;**345**:469-473. DOI: 10.1016/j.carres.2009.12.010

[14] Tipson S. Infrared Spectroscopy of Carbohydrates. National Bureau of Standards Monograph. Vol. 110. Washington, DC; 1968. https://digital.library.unt.edu/ark:/67531/metadc70397/

[15] Aspinall GO. The Polysaccharides. New York: Academic Press; 1982. pp. 172-184

[16] Wikanta T, Nasution RR, Lestari R. Effect of κ-carrageenan and ι-carrageenan feeding on the reduction of hyperglicemic rat blood glucose level. Jurnal Pascapanen dan Bioteknologi Kelautan dan Perikanan. 2008;**3**:131-138. DOI: 10.15578/jpbkp.v3i2.18

[17] Elekofehinti OO, Kamdem JP, Kade IJ, Rocha JBT, Adanlawo IG. Hypoglycemic, antiperoxidative and antihyperlipidemic effects of saponins from *Solanum anguivi* lam. Fruits in alloxan-induced diabetic rats. South African Journal of Botany. 2013;**88**:56-61. DOI: 0.1016/j.sajb.2013.04.010

[18] Gandhi GR, Sasikumar P. Antidiabetic effect of *Merremia emarginata* Burn. F. In strepto-zotocin induced diabetic rats. Asian Pacific Journal of Tropical Biomedicine; **2**:281-286. DOI: 10.1016/S2221-1691(12)60023-9

[19] Etuk EU. Animals models for studying diabetes mellitus. Agriculture Biology Journal of North America. 2010;**1**:130-134. https://scihub.org/ABJNA/PDF/2010/2/1-2-130-134.pdf

[20] Szkudelski T. The mechanism of alloxan and streptozotocin action in B-cells of the rat pancreas. Physiologycal Research. 2001;**50**:537-546. http://www.biomed.cas.cz/physi-olres/pdf/50/50_537.pdf

[21] Giugliano D, Ceriello A, Exposito K. Glucose metabolism and hyperglycemia. The American Journal of Clinical Nutrition. 2008;**87**:217S-222S. DOI: 10.1093/ajcn/87.1.217S

[22] Nelson RW, Duesberg CA, Ford SL, Feldman EC, Davenport DJ, Neal L. Effect of dietary insoluble fiber on control of glycemia in dogs with naturally acquired diabetes mellitus. Journal of the American Veterinary Medical Association. 1998;**212**:380-386. https://www.ncbi.nlm.nih.gov/pubmed/9470048

[23] Nelson RW, Scott-Moncrieff JC, Feldma EC, DeVries-Concannon SE, Kass PH, Davenport DJ, Kiernan CT, Neal LA. Effect of dietary insoluble fiber on control of glycemia in cats with naturally acquired diabetes mellitus. Journal of the American Veterinary Medical Association. 2000;**216**:1082-1088. DOI: 10.2460/javma.2000.216.1082

[24] Meyer U, Gressner AM. Endocrine regulation of energy metabolism: Review of patho-biochemical and clinical chemical aspects of leptin, ghrelin, adiponectin, and resistin. Clinical Chemistry. 2004;**50**:1511-1525. DOI: 10.1373/clinchem.2004.032482

[25] Wolf BW, Lai CS, Kipnes MS, Ataya DG, Wheeler KB, Zinker BA, Garleb KA, Firkins JL. Glycemic and insulinemic responses of nondiabetic healthy adult subjects to an experimental acid-induced viscosity complex incorporated into a glucose beverage. Nutrition. 2002;**18**:621-627. DOI: 10.1016/S0899-9007(02)00750-5

[26] Torsdottir I, Alpsten M, Holm G, Sandberg AS, Tölli J. A small dose of soluble algi-nate-fiber affects postprandial glycemia and gastric emptying in humans with diabetes. Journal of Nutrition. 1991;**121**:795-799. https://www.ncbi.nlm.nih.gov/pubmed/1851824

[27] Moree SS, Kavishankarb GB, Rajeshaa J. Antidiabetic effect of secoisolariciresinol diglu-coside in streptozotocin-induced diabetic rats. Phytomedicine. 2013;**20**:237-245. DOI: 10.1016/j.phymed.2012.11.011

[28] Weiss RB. Streptozocin: A review of its pharmacology, efficacy, and toxicity. Cancer Treatment Report. 1982;**66**:427-438. https://www.ncbi.nlm.nih.gov/pubmed/6277485

[29] Nugroho AE. Hewan percobaan diabetes mellitus: patologi dan mekanisme aksi diabe-togenik [Animal models of diabetes mellitus: Pathology and mechanism of some diabe-togenics]. Biodiversitas. 2006;**7**:378-382. DOI: 10.13057/biodiv/d070415

[30] Yavorska N. Sodium alginate – A potential tool for weight management: Effect on subjective appetite, food intake, and glycemic and insulin regulation. Journal of Undergraduate Life Sciences. 2012;**6**:66-69. http://juls.ca/717-2/

[31] Wikanta T, Damayanti R, Rahayu L. Effect of κ-carrageenan and ι-carrageenan feeding on the reduction of hyperglicemic rat blood glucose level [Pengaruh pemberian k-karagenan dan i-karagenan terhadap penurunan kadar glukosa darah tikus hiperglikemia]. Jurnal Pascapanen dan Bioteknologi Kelautan dan Perikanan. 2008;**3**:131-138. DOI: 10.15578/jpbkp.v3i2.18

[32] Tomkin GH. Targets for intervention in dyslipidemia in diabetes. Diabetes Care. 2008;**31**:S241-S248. DOI: https://doi.org/10.2337/dc08-s260

[33] Suzuki T, Nakai K, Yoshie Y, Shirai T, Hirano T. Effect of sodium alginates rich in guluronic and mannuronic acids on cholesterol levels and digestive organs of high-cholesterol-fed rats. Nippon Suisan Gakkaishi. 1993;**59**:545-551. DOI: https://doi.org/10.2331/suisan.59.545

[34] Ren D, Noda H, Amano H, Nishino T, Nishizawa K. Study on antihypertensive and antihyperlipidemic effects of marine algae. Fisheries Science. 1994;**60**:83-88. DOI: https://doi.org/10.2331/fishsci.60.83

[35] Wikanta T, Nasution RR, Rahayu L. Pengaruh pemberian natrium alginat terhadap penurunan kadar kolesterol total darah dan bobot badan tikus. Jurnal Penelitian Perikanan Indonesia. 2003;**9**:23-31. DOI: 10.15578/jppi.9.5.2003.23-31

[36] Rohman MS. Patogenesis dan terapi sindroma metabolik. Jurnal Kardiologi Indonesia. 2007;**28**:160-168. http://www.jki.or.id/index.php/jki/issue/view/33

[37] Goldberg IJ. Diabetic dyslipidemia: Causes and consequences. The Journal of Clinical and Endocrinology and Metabolism. 2001;**86**:965-971. DOI: 10.1210/jcem.86.3.7304

[38] Rachmat R, Rasyid A. Aktivitas Antihypercholesterolemia Alginat yang Diisolasi dari Sargassum carssifolium. Prosiding Pra Kongres Ilmu Pengetahuan Nasional VII. Forum Komunikasi I. Ikatan Fikologi Indonesia; 1999. pp. 111-118

[39] Baraas F. Kardiovaskuler Molekuler. Jakarta: Yayasan Kardia Iqratama; 2006

[40] Paxman JR, Richardson JC, Dettmar PW, Corfe BM. Alginate reduces the increased uptake of cholesterol and glucose in overweight male subjects: A pilot study. Nutrition Research. 2008;**28**:501-505. DOI: 10.1016/j.nutres.2008.05.008

[41] Pejic RN, Lee DT. Hypertriglyceridemia. Journal of the American Board of Family Medicine. 2006;**19**:310-316. DOI: 10.3122/jabfm.19.3.310

[42] Pujar A, Kumar A, Sridhar M, Kulkarni SV. An interesting case of hypertriglyceridemic pancreatitis. Journal of Clinical and Diagnostic Research. 2013;**7**:1169-1171. DOI: 10.7860/JCDR/2013/5500.3080

[43] Golstein P, Kroemer G. Cell death by necrosis: Towards a molecular definition. Trends in Biochemical Sciences. 2006;**32**:37-43. DOI: 10.1016/j.tibs.2006.11.001

[44] Sandberg AA, Philip DH. Interactions of exocrine and endocrine pancreatic diseases. Journal Pancreas. 2008;**9**:541-575. http://www.joplink.net/prev/200807/200807_07.pdf

[45] Holemans K, Bree RV, Verhaeghe J, Meurrens K, Assche AV. Maternal semistarvation and streptozotocin-diabetes in rats have different effects on the in vivo glucose uptake by peripheral tissues in their female adult offspring. The Journal of Nutrition. 1997;**127**:1371-1376. DOI: 10.1093/jn/127.7.1371

[46] Elias D, Prigozin H, Polak N, Rapoport M, Lohse AW, Cohen IR. Autoimmune diabetes induced by the β-cell toxin STZ: Immunity to the 60-kDa heat shock protein and insulin. Diabetes. 1994;**43**:992-998. DOI: 10.2337/diab.43.8.992

[47] Ikebukuro K, Adachi Y, Yamada Y, Fujimoto S, Seino Y, Oyaizu H. Treatment of streptozotocin-induced diabetes mellitus by transplantation of islet cells plus bone marrow cells via portal vein in rats. Transplantation. 2002;**73**:512-528. DOI: 10.1097/00007890-200202270-00004

Chapter 4

Tilapia (*Oreochromis aureus*) Collagen for Medical Biomaterials

David R. Valenzuela-Rojo,
Jaime López-Cervantes and
Dalia I. Sánchez-Machado

Additional information is available at the end of the chapter

http://dx.doi.org/10.5772/intechopen.77051

Abstract

Collagen is a natural polymer widely used in pharmaceutical products and nutritional supplement due to its biocompatibility and biodegradability. Collagen is a fibrous protein that supports various tissues, and its primary structure is formed by repeated units of glycine-proline-hydroxyproline. Traditional sources of collagen, such as bovine and pig skins or chicken waste, limit their use due to the dangers of animal-borne diseases. Thus, marine animals are an alternative for the extraction of collagen. The common name of *Oreochromis aureus* is tilapia, widely cultivated for sale as frozen fillets. During its processing, a large amount of collagen-rich wastes are generated. Therefore, the objective of this book chapter is to prove the potential of tilapia skin as an alternative source of collagen for the elaboration of biomaterials. Additionally to the literature review, experimental results of the extraction and characterization of tilapia skin collagen for use in medical dressings are presented.

Keywords: acid soluble collagen, marine byproducts, fish skin, valorization, biomimetic materials

1. Introduction

Collagen is widely used in the manufacture of medical materials due to its availability, versatility, compatibility, and degradability in advantage with other biomimetic biomaterials [1]. Traditionally, most of the collagen is extracted from the skins of cattle and pigs or chicken waste. However, these sources limit their applications due to the religious beliefs of some consumers and because of the risk of diseases such as bovine spongiform encephalopathy and aphthous fever disease [2]. During the production of commercial fish products, byproducts such as skin, bones, and scales rich in collagen are generated [3].

The isolation of collagen from byproducts from fish would reduce the environmental impact generated during its decomposition by providing an added value to these wastes. The collagen isolated from fish is easier to digest and absorb than that of terrestrial origin, due to the different hydroxyproline contents and a lower denaturation temperature [4]. However, the information reported on the characteristics and methods of extracting collagen is still not enough. Tilapia (*Oreochromis aureus*) is one of the main groups of fish grown and sold as whole frozen fish and frozen and fresh fillet. Tilapia skin is a byproduct that contains about 27.8% collagen and can be used for the extraction of collagen increasing the economic value of these industries [5]. Muyonga et al. have reported the extraction of collagen in acid medium of Nile perch [6]. Liu et al. established that the concentration of acetic acid and temperature of extraction of collagen have effects on the properties of collagen [7].

Tilapia skin collagen can be used to develop healing biomaterials and cosmetics or as food supplement due to its biological properties. In the literary review of this book chapter, the residues of the industrialization of tilapia (*O. aureus*) are presented as sources for the extraction of collagen. Basically, the collagen extraction and characterization methods are detailed with a theoretical and practical perspective for the fishing industries in order to generate biological products with high commercial value. The main purpose of this chapter is show the potential of soluble collagen in acid from tilapia skin as a biomaterial with medical applications.

2. Collagen

Collagen is the most abundant protein in vertebrates and contributes significantly to the hardness of connective tissues such as the skin, bones, cartilage, tendons, and blood vessels [8]. Due to its biological compatibility, biodegradability, low cytotoxicity, structural support, and hemostatic activity, collagen has been widely used in the food, pharmaceutical, cosmetological, and biomedical industries. Currently, at least 29 types of collagen have been identified, and these are classified according to their structure as fibrous, nonfibrous, and micro-fibrillary. Type I collagen is the most common, present mainly in the skin, tendons, and bones, while type II collagen is found in cartilage tissues, and type III collagen depends on the age of the tissue. The other types of collagen are only in very small amounts and in specific organs [9].

2.1. Chemical structure

It has been reported that collagen is composed of glycine (33%), proline (12%), alanine (11%), and hydroxyproline (10%) but is deficient in essential amino acids such as lysine and tryptophan [10]. The collagen monomer is called tropocollagen and is a cylindrical protein of 15 Å in diameter and 3000 Å in length, which is formed by three polypeptide chains or α chains of molecular weight of 100,000 daltons each [11]. The three chains are linked together through cross intermolecular bonds giving rigidity to the structure and poor solubility. Collagen is produced from the interaction of tropocollagen molecules [12].

The amino acid sequence of the polypeptide chain is characterized by the repetition of Gly-X-Y, where the X and Y positions are normally occupied by proline and hydroxyproline, respectively [13]. The glycine of the polypeptide chain becomes the helical center of the molecule, because it only has a side chain of hydrogen. This structure is reinforced by pyrrolidine rings from proline and hydroxyproline [14].

In terms of amino acid composition and biocompatibility, the aquatic-type collagen has properties similar to mammalian sources. In both sources of collagen, glycine is one of the most abundant amino acids representing about 30%, while hydroxyproline is estimated in the range of 35–48% of the total amino acid content [15]. Glycine and hydroxyproline are important for the formation of the structure and characteristic stability of collagen. However, some differences in the amino acid content between aquatic species have been found. Swatschek et al. have reported that collagen isolated from marine sponges (*Chondrosia reniformis*) showed a content of glycine and hydroxyproline of 18.9 and 40%, respectively [16]. However, Silva et al. [17] reported values of 31.6% glycine and 47.36% hydroxyproline for collagen isolated from the same species. This discrepancy is due to some structural and chemical differences between the sources. In addition, glycoproteins in marine tissues generate differences in amino acid content, because they are associated with collagen and appear as impurities affecting the purity of collagen [16].

Compared with mammalian sources, aquatic collagen has a lower viscosity and thermal stability due to the difference between the content of proline and hydroxyproline in its structure [18]. Gómez-Guillén et al. [19] have reported that the content of proline and hydroxyproline in collagen isolated from cod and squid skin is similar. The structure of the collagen can be denatured under conditions of high temperatures forming gelatin. Gelatin is a mixture of peptides and proteins partially hydrolyzed from collagen molecules [20]. Gelatin is classified as the denatured form of native collagen with lower molecular weights than its predecessor [18]. Compared to gelatin, collagen exhibits superior characteristics such as greater enthalpy, structuring of fibrous networks, basic isoelectric point and high resistance to protease hydrolysis, greater mechanical resistance, and reversible extensibility, whereas the gelatin shows gel formation, great thermal stability, and unique rheological properties [21].

2.2. Natural sources

Collagen is one of the most common proteins in multicellular organisms and, due to its fibrous nature, provides structural rigidity to the connective tissues and internal organs. The waste generated during industrial food processing is the main source of collagen. Traditionally, the skin, bones, and cartilage of bovines and pigs are the materials for their isolation [22]. However, the application of these sources has limitations by certain religious and ethnic groups [18]. In addition, disease transmission is a probability, such as bovine spongiform encephalopathy and aphthous fever disease [2]. At the present, there is a growing interest in the valuation of industrial byproducts of aquatic origin as an alternative to conventional raw materials for the production of collagen.

During the industrial processing of fishery products, about 75% of the initial weight is discarded as waste, which are sources of collagen that can increase the profitability of the industry [23]. The skin, scales, fins, and bones of marine and freshwater fish are recognized as potential sources of collagen for the production of food, cosmetics, and pharmaceutical products [5]. Unlike collagen from terrestrial sources, that of aquatic origin is immune to diseases; of faster absorption due to its denaturation temperature, metabolic compatibility, molecular weight; and of low inflammatory response.

2.3. Biological properties

For the development of biomaterials, there are several desirable properties such as biodegradation, hemostasis, cell proliferation, immunogenic, and biocompatibility [24]. In addition to possessing all of them, collagen is resistant to traction with minimal extensibility. However, this depends on the amount and type of collagen, as well as the interaction with different glycoproteins and proteoglycans.

Ulery et al. collagen has an important function in coagulation, from the conversion of fibrinogen into fibrin, capturing platelets and forming clots [25]. Collagen as a hemostatic agent works as a mechanical and protective blockage from environmental factors, preserves epithelial cells, and increases the production and permeation of fibroblasts at the collagen-wound interface. Hemostasis is generated in less than 5 min; the process begins with the adhesion of platelets around the structure of collagen, activating a coagulation factor and epithelization. In the structure of collagen, the amino acid arginine participates in healing, decreasing stress in tissues and increasing the interaction of collagen with platelets. In addition, many studies show the influence of collagen on cell migration, adhesion, and differentiation [26, 27].

The biodegradation of collagen is carried out by the action of metalloproteinase enzymes. Type I collagen is hydrolyzed by collagenase (MMP-1) which has the ability to break down the triple helix of collagen, resistant to most proteases. Then, the resulting collagen fragments are degraded by gelatinases and other proteases [28].

The biocompatibility of collagen is a valuable aspect in most biomaterials, because when transplanting collagen to a damaged tissue, it does not leave strange residues for the organism during its degradation [24].

Collagen has low toxicity and a low probability to trigger an immune response, making it a suitable material for use as a biomaterial in the medical industry. However, a risk exists for people susceptible to collagen, which is why a serological test is currently available to identify people who are allergic to this protein [1, 25].

2.4. Methods of extraction of collagen of aquatic origin

The methodology for the extraction of collagen includes steps such as pretreatment of the raw material and isolation of collagen. The fish byproducts as collagen sources need a previous cleaning and size reduction, in order to facilitate the elimination of impurities and ensure the maximum collagen extraction. In addition, it is important to classify the waste generated by industrial processing such as the skin, scales, fins, and bones since they have characteristic elements that require different methodologies for the isolation of collagen [29]. **Table 1** shows various processing residues of aquatic species as sources of collagen.

For the removal of non-collagen proteins and pigments, an alkaline treatment is recommended, and diluted NaOH has been used the most because the performance of the collagen is not affected. However, increasing the NaOH concentration induces significant collagen losses [7]. Skierka et al. [34] propose using NaCl to remove non-collagen proteins from cod skin, but NaOH has shown greater efficacy than NaCl. The most effective method for the removal of fat from fish skin is with butyl alcohol [35]. The demineralization of scales and

Source	Species	Yield (%)	References
Scale	*Hypophthalmichthys nobilis*	PSC 2.7	[8]
Scale	*Ctenopharyngodon idellus*	ACS 16.1	[30]
Scale	*Oreochromis niloticus*	PSC 14.9	[5]
Bone	*Magalaspis cordyla*	ACS 30.5	[31]
		PSC 27.6	
Bone	*Otolithes ruber*	ACS 45.1	[31]
		PSC 48.6	
Skin	*Labeo rohita*	ACS 46.13	[23]
Skin	*Oreochromis niloticus*	ACS 39.4	[5]
Skin	*Misgurnus anguillicaudatus*	ACS 22.42	[32]
		PSC 27.32	
Skin	*Oreochromis niloticus*	ACS 27.2	[33]

Note: ACS, acid soluble collagen; PCS, pepsin soluble collagen.

Table 1. Potential sources of collagen of aquatic origin.

bones has been carried out with citric acid, HCl, and ethylenediaminetetraacetic acid (EDTA), achieving efficiencies greater than 90%. However, the selection of the chemical agent is of great importance due to the loss of collagen with the use of these acids [36].

After the removal of non-collagen proteins, demineralization, and defatting of the byproducts, the collagen is extracted. The properties and performance of collagen depend on the extraction procedures; therefore, it is important to establish the right conditions [23]. Collagen receives its name according to the methodology applied for its extraction.

Collagen soluble in salt can be isolated with solutions of NaCl and NaOH, although this method is not the most popular. In reference to Wang et al. [30], the authors isolated and characterized collagen from Amur sturgeon (*Acipenser schrenckii*) by several methods, with 0.45 M NaCl at pH 7.5 for 24 h, 0.5 M acetic acid, and hydrolysis with pepsin, and reported yields of 4.55, 37.42, and 52.80%, respectively.

Collagen soluble in acid is extracted with an acidic solution. This is the traditional method for extracting collagen, and hydrochloric, citric, acetic, and lactic acids have been used. Skierka et al. have reported higher performance when using organic acids such as acetic and lactic acid [34]. Collagen soluble in acetic acid has been isolated from different marine byproducts such as the skin [23], scales, bones [30], and fish fins [8]. To increase the collagen concentration, the supernatant is precipitated with NaCl. Then the precipitate is centrifuged and redissolved in 0.5 M acetic acid. Finally, it is dialyzed in acetic acid and lyophilized [32]. However, the purification costs and the process times are high, limiting their use in industrial processes.

To increase the yield of the isolated collagen, it is necessary to hydrolyze the collagen from insolubilities by means of enzymes. Enzymes such as trypsin, pancreatin, ficin, bromelain, papain, and pepsin have been used for this process. The latter has been the most commonly used in marine byproducts [37]. Pepsin-soluble collagen obtained by hydrolysis with pepsin is called PCS or atelocollagen. This treatment separates the peptides specifically in the telopeptide region of collagen; these are non-helical ends, the hydrolysis being a more efficient process that decreases the toxicity caused by the telopeptides [29]. After hydrolysis, the collagen is centrifuged and dissolved in 0.5 M acetic acid, dialyzed, and lyophilized [38].

Kim et al. have reported that the extraction of collagen with ultrasound increases the performance, even when a lower concentration of acetic acid is used over than the traditional method [39]. However, Ran and Wang state that the ultrasound extraction method could break the hydrogen bonds in the collagen chains, causing denaturation of the protein and the enzyme used for its isolation [40]. In reference to Yu et al., various enzymes were used for the extraction of collagen, finding that the activity of papain was inhibited by structural changes in its activity, while pepsin did not change [41]. In addition, they reported that the application of ultrasound for long periods of time causes a rise in temperature leading to the denaturation of collagen.

Huang et al. have proposed an extrusion-hydro-extraction method for the extraction of collagen from tilapia (*Oreochromis* sp.) scales. Basically, the process consists of three main stages, preconditioning (citric or acetic acid), extrusion (360 rpm at 135°C), and hydro-extraction (25 and 50°C). The precise combination between heat and mechanical strength favors the extraction of collagen. The collagen isolated during this procedure was identified as type I, and the physicochemical properties of collagen revealed that it can be used for food, cosmetic, and medical applications [42].

3. Biomedical applications

Collagen, due to its multiple properties, has numerous applications in the food, medical, dental, cosmetic, and pharmaceutical industries. In addition, depending on its use, it can be processed in a wide variety of ways such as powders, injectable solutions, films, sponges, and hydrogels [29].

3.1. Food supplement

Collagen has been mixed with a wide variety of products and beverages, which is why collagen-based food supplements are the most common on the market. Collagen synthesis decreases with age, and the tissues lose flexibility and thickness, reason why collagen supplements are mainly used for skin care. In addition, the consumption of collagen increases the gain of muscle mass, decreases the recovery time of the muscles, and helps to rebuild damaged joints [21].

In the food industry, collagen is used as an additive for the improvement of the rheological properties of charcuterie and as a guarantee of the presence of nutritive fibers of animal origin [43]. Collagen functions as a barrier that controls the migration of oxygen, providing permeability to water vapor and prolonging the shelf life of food [21]. These films or edible coatings can be applied by wrapping, dipping, brushing, or spraying the food. Currently the functional beverage sector has increased its demand, requiring the implementation of new ingredients to meet the various nutritional needs. According to Bilek and Bayram, they have developed functional drinks from orange, apple, and grape juice, mixed with hydrolyzed collagen. Due to its characteristic amino acid content, hydrolyzed collagen is a promising ingredient in the food industry, specifically for the production of beverages because it increases the protein content and bioavailability of collagen [44].

3.2. Cosmetology

Collagen is one of the main components of the skin and is responsible for its appearance and physical condition. The collagen present in the skin decreases with age and with prolonged exposure to ultraviolet irradiation. Therefore, it is important to compensate these losses by consuming products rich in this protein for cosmetic and pharmaceutical purposes [45].

Compared to high molecular weight collagen, cosmetic formulations use hydrolyzed collagen. This is due to its superior solubility at neutral pH and ease of penetration into the dermis and because it acts as a water-binding agent inside the skin. Thus, its addition in cosmetic creams increases its wetting capacity, providing a moisturizing, softening, and glowing effect to the skin [12].

Collagens of aquatic origin have potential as excellent ingredients for the cosmetic industry, reason why creams and gels have been developed with a high moisturizing action, anti-wrinkles, and UV protectors [29].

3.3. Healing materials

Collagen of aquatic origin has proven its use an alternative material in the manufacture of medical dressings such as sponges and membranes for the treatment of wounds. It is also included in biomaterials of medical uses for ophthalmology, bone substitutes, and gels for the administration of drugs [12].

3.3.1. Scaffolds

Collagen matrices have the ability to absorb large amounts of exudates from wounds. This favors the formation of a biodegradable gel or sheet on the surface of the wound that maintains a humid environment, promotes healing, and provides protection against external mechanical forces [46]. Collagen sponges possess adequate characteristics for tissue regeneration due to their high porosity, permeability, low toxicity, cell adhesion, and biocompatibility [47].

For the production of collagen sponges, aqueous solutions are prepared and finally lyophilized. The porosity of the sponge is controlled by varying the rate of freezing prior to lyophilization and the concentration of collagen in the solutions [28].

Chandika et al. developed collagen sponges from the skin of Japanese halibut (*Paralichthys Olivaceus*) in combination with alginate and chitosan; these exhibited a porous structure, high capacity of swelling, and biodegradation suitable for its application in tissues [48]. Cheng et al. prepared biomaterials of type I collagen isolated from jellyfish (*Rhopilema esculentum*) for the treatment of wounds. In vivo studies indicate that collagen sponges exhibit rapid hemostatic properties due to their ability to absorb liquids [49].

Biomaterial	Compound	Source of collagen	References
Sponge	Collagen/alginate/chitosan cross-linking with glutaraldehyde	Skin japanese halibut (P. Olivaceus)	[48]
Sponge	Collagen	Jellyfish (Rhopilema esculentum)	[49]
Sponge	Chitosan/collagen/hydroxypatite	Marine sponge (Ircinia fusca)	[51]
Membrane	Collagen/hydroxypropyl methylcellulose E15	Skin marine eel (Evenchelys macrura)	[52]
Membrane	Collagen/glycerol	Skin silver carp (Hypophthalmichthys molitrix)	[53]

Table 2. Healing materials based on aquatic collagen.

The collagen content in a sponge increases its porosity and biodegradation. This effect was described by Ullah et al. [50] in collagen sponges isolated from tilapia (*Oreochromis* sp.) scales mixed with chitosan and glycerin for the regeneration of damaged tissues. **Table 2** shows various biomedical materials based on aquatic collagen.

3.3.2. Membranes

Collagen membranes or collagen films have an average thickness of 0.01–0.5 mm and are formed by drying aqueous solutions by aeration. The membranes provide a protective barrier from the environment, are resistant to handling during their application, and can be steril-ized. Collagen membranes have been used in the treatment of skin lesions, corneal tissue, and reinforcement in compromised tissues and as a vehicle for encapsulated drugs with slow release [12, 28].

Veeruraj et al. [52] developed collagen membranes isolated from *Evenchelys macrura* for the administration of standard commercial medicines such as ampicillin. However, Perumal et al. prepared collagen-fucoidan membranes as a substrate for the growth of fibroblasts [54]. Liu et al. modified the collagen membranes by crosslinking with alginate to improve thermal sta-bility and mechanical properties and decrease the rate of biodegradation [55].

3.3.3. Ophthalmologic inserts

Collagen being biocompatible and biodegradable has been used in ophthalmology, as grafts for the replacement of corneas, suture material, and ocular protective films. Raiskup et al. applied a corneal collagen implant with riboflavin for the effective treatment of progressive keratoconus, reducing astigmatism and corneal distortion [56]. In reference to Zhang et al. [3], the authors proposed a type I collagen sponge obtained from rat tail as a substitute for corneal tissue. Long et al. evaluated collagen-hydroxypropyl methylcellulose membranes, which had optical properties equivalent to human corneas [57].

3.3.4. Bone substitutes

In comparison with other biomolecules, collagen has the great advantage of forming biomaterials with network and porous structures, with a mechano-elastic behavior suitable for various biomedical applications [58]. Collagen participates in the treatment of the cartilage and bone by cell proliferation and is able to regenerate tissue damaged by injury or wear [59]. The mechanical properties of bone substitutes are very important; collagen can be used in combination with other biomaterials such as hydroxyapatite, calcium phosphate, chitosan, and alginate, giving them different mechanical and biological properties than the original polymer [51].

According to Aravamudhan et al. [60], it has been reported that collagen-cellulose nanofibers with microporous structure are alternative candidates for the repair and regeneration of bone defects, increasing the bone density and quality of mice. Wahl et al. have reported a collagen-hydroxyapatite compound that has the potential to mimic and replace bone fragments of the skeleton [61]. On the other hand, Murphy et al. [62] established an optimum porosity of 325 μm in collagen-glycosaminoglycan sponges for its application in bone regeneration, increasing cell adhesion and filtration, in comparison with other pore sizes.

3.3.5. Drug releasers

Collagen gels can act as a matrix for the administration of drugs due to their properties such as fluidity, easy handling, and biocompatibility. According to Calejo et al. [63], polymer matrix from jellyfish collagen for the supply of proteins demonstrated the potential of aquatic collagen in the controlled release of drugs. On the other hand, Dinescu et al. reported collagen-sericin sponges for the supply of hyaluronic acid and chondroitin sulfate, which presented biological properties such as viability and cell proliferation [64]. Langasco et al. have reported that in vitro studies using collagen sponges with L-cysteine hydrochloride favor the slow release of drugs [65].

4. Extraction of tilapia skin collagen

4.1. Tilapia (*Oreochromis aureus*)

The tilapia species are native to Africa and the Middle East, among the most common cultivated species is the genus *Oreochromis*. These include the Nile tilapia (*Oreochromis niloticus*), blue tilapia (*Oreochromis aureus*), and tilapia from Mozambique (*Oreochromis mossambicus*); these species can crossbreed producing hybrids available for cultivation. Tilapia has been cultivated in almost all countries of tropical climate and subtropical regions. This species is of global importance in aquaculture due to its ability to adapt to different conditions and high reproductive rate [66].

Tilapia has the ability to grow in wide ranges of salinity, from freshwater to seawater, and can tolerate acid (pH 5) and alkaline (pH 9) systems, low levels of oxygen (<2 mg/l), and high levels of ammonia (50 mg/l) [67]. Depending on the species, tilapia can be herbivorous or omnivorous, feeding on a wide variety of foods, including periphyton, phytoplankton, zoo-plankton, and balanced feed [68].

Tilapia is usually marketed in fresh-chilled form and frozen as whole or fillet. During the industrial processing of the fillets, large amounts of waste (skin, scales, and bones) rich in bioactive biomolecules are produced. These byproducts can be used for the extraction of collagen, which decreases the environmental impact and increasing the economic value of the company [5].

4.2. Preparation of tilapia skin

The skins of *Oreochromis aureus* from industrial processing were peeled and cut into pieces smaller than 2 cm and then stored frozen until use. For the removal of non-collagen proteins, the skins were immersed in a 0.1 N NaOH solution for 48 h, changing the solution every 24 h according to the method described by Zeng et al. [5]. Then, the skins were defatted with 10% butyl alcohol for 48 h with a change of solution every 24 h. Subsequently, the skins were washed with distilled water for waste disposal and dried for storage.

4.3. Isolation of acid-soluble collagen

The previously treated tilapia skins were submerged in 0.5 M acetic acid (1:50, w/v) for 3 days with gentle agitation, according to the method of Singh et al. [69] with some modifications. The mixture was then filtered through a 1 mm mesh, and the filtrate was lyophilized at −50°C with a maximum vacuum of 0.5 mBar. For the purification of the collagen, the lyophilized sample was resuspended in 0.5 acetic acid (1:30, w/v) for 2 days; then the solution was filtered through a mesh with a pore less than 1 mm. Finally, the filtrate was lyophilized and stored as collagen soluble in pure acid. The proximal composition was determined according to the AOAC [70] methods for moisture, ash, and proteins (Kjeldahl factor 6.25).

The process of extracting collagen from tilapia (*Oreochromis aureus*) skin is shown in **Figure 1**. The collagen extracted from tilapia skin has a uniform whitish color without any impurities. The average moisture content of the lyophilized collagen was 5.46 ± 0.49%, which determines its storage stability and influences its nutritional characteristics. In reference to the inorganic fraction, it was found that the ash content was 0.55 ± 0.02%; this indicates effectiveness of the method to eliminate minerals. Likewise, the average protein content was 87.54 ± 3.61%.

The yield of lyophilized collagen was estimated based on the weight of dry tilapia skins. Therefore, the yield of acid-soluble collagen was 11.37 ± 0.88% on a dry basis. The yield was lower than that reported by Chen et al. [33] with 27.2%. These differences may be due to the state of maturity of the fish, environmental conditions of its development, as well as the extraction method.

Also, it has been reported that collagen is not completely solubilized during extraction with acetic acid. Tamilmozhi et al. [71] reported trials with skins of *Istiophorus platypterus*. Hickman et al. established that molecules that solubilize in acid medium are monomeric [72].

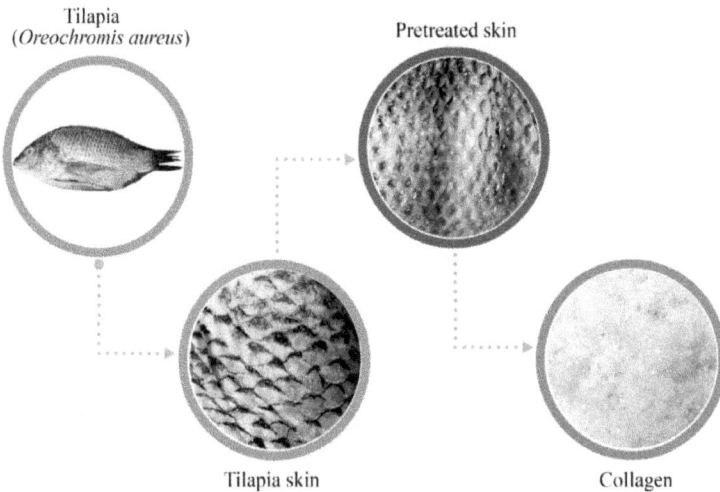

Figure 1. Diagram of extraction of collagen from tilapia (*Oreochromis aureus*) skin.

4.4. Characterization of acid-soluble collagen

4.4.1. Fourier transform infrared spectroscopy

For the identification of collagen, FTIR spectra were recorded (Thermo Scientific FTIR, model Nicolet 5s Madison, WI, USA). The equipment consists of an ATR ID3 accessory for germanium crystal reflection. The spectral resolution was 4 cm^{-1}, and 64 scans were obtained in the range of 600–4000 cm^{-1}.

The IR spectra showed the five characteristic absorption bands of collagen (amide A, amide B, and amides I, II, and III) suggesting the composition of amino acids and a high portion of proline and hydroxyproline in the collagen molecule [73]. Amide A is located at 3301 cm^{-1}, and it is associated with the N–H groups involved in the hydrogen bonds of the peptide chain, while amide B was measured at 3066 cm^{-1}, and it is related to the asymmetric stretching of C–H [32]. The peak corresponding to amine I was found at 1642 cm^{-1}; this is presented with a characteristic wavelength range of 1600–1700 cm^{-1} associated with the stretching vibrations in the C=O group, being an indication of the secondary structure of the peptide [31]. Amide II was measured at 1545 cm^{-1}, attributed to the stretching of the carbonyl group coupled to a carboxyl group. Amine III was measured at 1233 cm^{-1}, established by the combination of the stretching vibration between C–N and the bending vibration of N–H. The values found during the study confirm the presence of a triple helical structure characteristic of collagen, all in accordance with results reported by Sun et al. for collagen extracted from tilapia (*Oreochromis niloticus*) skin [73]. The FTIR spectrum of the acid-soluble collagen is shown in **Figure 2**.

4.4.2. UV absorption spectroscopy

The UV absorption spectra of the collagen isolates were measured in a UV spectrometer (ThermoScientific, Genesys 10S UV–Vis, USA), by the procedure described by Liu et al. [55]

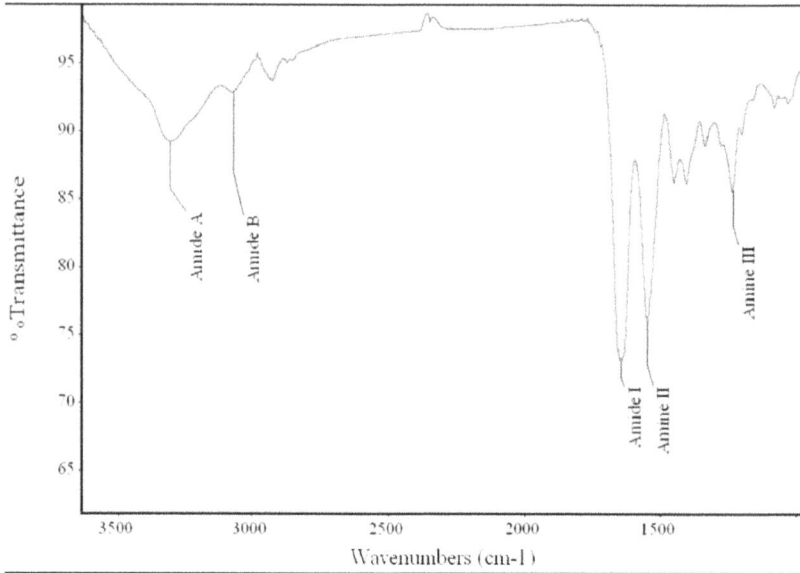

Figure 2. FTIR spectrum of collagen from tilapia (*Oreochromis aureus*) skin.

with some modifications. The lyophilized samples were dissolved in 0.5 M acetic acid to obtain a concentration of 1 mg/ml and centrifuged at 5000 rpm for 15 min. The absorbance of the supernatant was measured at different lengths in the range of 200–400 nm, with a range of 1 nm to record its characteristic spectrum and maximum absorbance.

The UV absorption of collagen is attributed mainly to the peptide bonds and side chains in its structure. The collagen isolated from the skin of tilapia showed a maximum absorption at 230 nm; this agrees with the characteristic absorbance of type I collagen [31]. They also coincide with different authors, who isolated collagen from different marine species [32, 35]. The UV–Vis spectrum of tilapia (*Oreochromis niloticus*) skin collagen is presented in **Figure 3A**.

4.4.3. Denaturing temperature of collagen

The denaturation temperature was determined according to the method of Sun et al. [73] based on the viscosity changes in relation to the temperature increase measured with an Ubbelohde viscometer. A solution of collagen at 1 mg/ml dissolved in 0.5 M acetic acid was prepared, the solution was loaded into the viscometer, and the temperature curves were recorded in the range of 10–45°C. The fractional viscosity for each specific temperature was calculated with the following equation: fractional viscosity = $(\eta_{sp}(T) - \eta_{sp}(45°C))/(\eta_{sp}(10°C) - \eta_{sp}(45°C))$, where η_{sp} is the specific viscosity, specific viscosity = $(t - t_0)/t_0 = \eta_r - 1$, where η_r is the relative viscosity, relative viscosity = t/t_0, where t is the time it takes for the collagen solutions to pass through the capillary of the Ubbelohde viscometer and t_0 is the time in which the solvent passes at the same temperature. The thermal denaturation curve was established by comparing the fractional viscosities and temperatures. The denaturing temperature is where the fractional viscosity is 0.5.

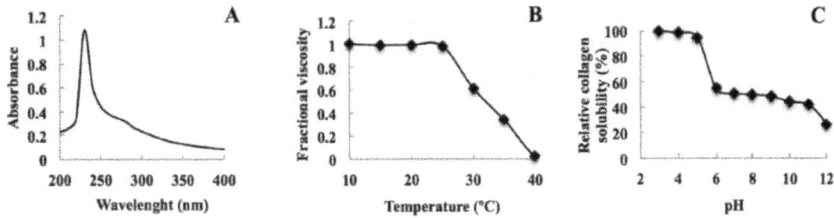

Figure 3. Characteristic UV spectrum (A), thermal denaturation curve (B), and solubility effect at different pH (C) collagen isolated from tilapia (*Oreochromis aureus*) skin.

The denaturation temperature found for acid-soluble collagen in tilapia skin was 32°C, similar to that reported by Sun et al. [73] for Nile tilapia skin, 35.2°C. However, for most marine species, these values range from 26 to 29°C [74] such as the following *Alaska pollack* (16.8°C), *Hypophthalmichthys molitrix* (29°C), and *Trachurus* (28.4°C) [6, 35, 74]. The connective tissue of the fish is continuously renewed, presenting a smaller amount of molecular crosslinks, which affects its thermal stability. In addition, the thermal stability of collagen correlates with the environmental temperatures present in the habitat of the species; this explains the difference between a higher Td of the collagen isolated from species of warm water compared to those of cold water, reason why subtropical and tropical fish present better thermal stability [75].

The denaturing temperature is the point where the collagen triple helix structure is deformed to a random spiral structure. This property is undesirable for the manufacture of biomaterials, because the denaturation drastically alters the physicochemical, biological, and mechanical properties of collagen. According to Fu et al. [76], this temperature can be measured by the viscosity changes caused by the heating of the collagen. **Figure 3B** shows the changes in the fractional viscosity with the increase in temperature for rehydrated collagen isolates in 0.5 M acetic acid from tilapia (*O. aureus*) skin.

The solubility of acid-soluble collagen was determined for different pH values (3–12) and for different concentrations of NaCl (0, 1, 2, 3, 5, and 6%) according to the method of Fu et al. [76] and Montero et al. [77]. All these solutions were centrifuged at 4000 g for 30 min, and the protein content of the supernatant was determined by the method of Bradford [78]. Finally, the relative solubility of collagen at different pH was calculated by comparison with the highest solubility presented at a given pH. Similarly, the relative solubility of collagen at different concentration of NaCl was calculated by comparison with the higher solubility at a given concentration of NaCl.

4.4.4. Effect of pH and concentration of NaCl on solubility

Collagen solutions from tilapia (*O. aureus*) skin showed a high solubility between the pH ranges of 3–5, while the maximum solubility was at pH 3. When increasing the pH to 6, the relative solubility of the collagen was 58.82 ± 4.04%; from pH 12 the solubility decreased to 32.35 ± 5.93%. According to Liu et al. [8], considerable decrease in solubility from pH 6 for collagen from different aquatic sources was reported. Chen et al. have reported a maximum solubilization at pH 3 and minimum at pH 7 for collagen of tilapia (*Oreochromis niloticus*) skin [33]. When the pH is equal to or close to the isoelectric point, the total charge of the protein molecules is close to zero and results in precipitation [79]. The effect of pH on the collagen solubility of tilapia (*O. aureus*) skin is shown in **Figure 3C**.

The collagen showed greater solubility in NaCl concentrations between 1 and 2%, while at 3% of NaCl, a minimal decrease was observed ($77.13 \pm 9.27\%$). When increasing the concentration of NaCl to 6%, the solubility decreased to $34.60 \pm 1.96\%$. The solubilities of shark (*Sphyrna lewini*) skin, bighead carp (*Hypophthalmichthys nobilis*), tilapia (*Oreochromis niloticus*), catfish (*Pangasianodon hypophthalmus*), and mackerel sardine (*Scomberomorus niphonius*) collagen decrease at concentrations greater than 2% NaCl [8, 34]. The decrease in collagen solubility with the increase in NaCl is due to the increase in hydrophobic interactions that favor the precipitation of proteins [80].

5. Conclusion

This literary review of collagen shows that this natural polymer possesses biological properties that benefit human health and justify its application as a medical biomaterial. Essentially, collagen is a biodegradable and biocompatible structural protein that has been used to enrich foods for athletes and novel cosmetics for skin care. Here, we propose a simple methodology for the extraction of collagen from the skin of tilapia (*Oreochromis aureus*) with properties similar to those of collagen from conventional sources.

Acknowledgements

The first author is grateful to CONACYT (477741). This research was financed under Project No. 248160 from CONACYT-PN2014 and by Project PROFAPI No. 2018-0010 from Instituto Tecnológico de Sonora.

Author details

David R. Valenzuela-Rojo, Jaime López-Cervantes and Dalia I. Sánchez-Machado*

*Address all correspondence to: dalia.sanchez@itson.edu.mx

Departamento de Biotecnología y Ciencias Alimentarias, Instituto Tecnológico de Sonora, Ciudad Obregón, Sonora, Mexico

References

[1] Parenteau-Bareil R, Gauvin R, Berthod F. Collagen-based biomaterials for tissue engineering applications. Materials. 2010;**3**(3):1863-1887. DOI: 10.3390/ma3031863

[2] Potaros T, Raksakulthai N. Characteristics of collagen from Nile Tilapia (*Oreochromis niloticus*) skin isolated by two different methods. Kasetsart Journal (Natural Science). 2009;**43**:584-593

[3]   Zhang Q, Wang Q, Lv S, Lu J, Jiang S, Regenstein J, Lin L. Comparison of collagen and gelatin extracted from the skins of Nile tilapia (*Oreochromis niloticus*) and channel catfish (*Ictalurus punctatus*). Food Bioscience. 2016;**13**:41-48. DOI: 10.1016/j.fbio.2015.12.005

[4]   El-Rashidy A, Gad A, Abu-Hussein A, Habib S, Badr N, Hashem A. Chemical and biological evaluation of Egyptian Nile Tilapia (*Oreochromis niloticas*) fish scale collagen. International Journal of Biological Macromolecules. 2015;**79**:618-626. DOI: 10.1016/j. ijbiomac.2015.05.019

[5]   Zeng S, Zhang C, Lin H, Yang P, Hong P, Jiang Z. Isolation and characterisation of acid-solubilised collagen from the skin of Nile tilapia (*Oreochromis niloticus*). Food Chemistry. 2009;**116**(4):879-883. DOI: 10.1016/j.foodchem.2009.03.038

[6]   Muyonga J, Cole C, Duodu K. Characterisation of acid soluble collagen from skins of young and adult Nile perch (*Lates niloticus*). Food Chemistry. 2004;**85**:8189. DOI: 10.1016/j.foodchem.2003.06.006

[7]   Liu D, Wei G, Li T, Hu J, Lu N, Joe M, Regenstein Z. Effects of alkaline pretreatments and acid extraction conditions on the acid-soluble collagen from grass carp (*Ctenopharyngodon idella*) skin. Food Chemistry. 2015;**172**:836-843. DOI: 10.1016/j.foodchem.2014.09.147

[8]   Liu D, Liang L, Regenstein J, Zhou P. Extraction and characterisation of pepsin-solubilised collagen from fins, scales, skins, bones and swim bladders of bighead carp (*Hypophthalmichthys nobilis*). Food Chemistry. 2012;**133**:1441-1448. DOI: 10.1016/j.foodchem.2012.02.032

[9]   Schmidt M, Dornelles R, Mello R, Kubota E, Mazutti M, Kempka AP, Demiate IM. Collagen extraction process. International Food Research Journal. 2016;**23**(3):913-922

[10]   Gálvez Mariscal A, Flores Argüello I, González SA. Proteínas. In: Gómez HC, Duarte EQ, editors. Química de los alimentos. 4th ed. Ciudad de México: Salvador Badui Dergal; 2006. pp. 119-135

[11]   Wong D. Química de los alimentos: mecanismos y teoría. 1st ed. Zaragoza: Acribia, S.A; 1994

[12]   Sionkowska A, Skrzyński S, Śmiechowski K, Kołodziejczak A. The review of versatile application of collagen. Polymers for Advanced Technologies. 2017;**28**:4-9. DOI: doi. org/10.1002/pat.3842

[13]   Shoulders M, Raines R. Collagen structure and stability. Annual Review of Biochemistry. 2009;**78**:929-958. DOI: 10.1146/annurev.biochem.77.032207.120833

[14]   Nagai T, Suzuki N, Nagashima T. Collagen from common minke whale (*Balaenoptera acutorostrata*) unesu. Food Chemistry. 2008;**111**(2):296-301. DOI: 10.1016/j.foodchem.2008.03.087

[15]   Yamada S, Yamamoto K, Ikeda T, Yanagiguchi K, Hayashi Y. Potency of fish collagen as a scaffold for regenerative medicine. BioMed Research International. 2014;**2014**:8, 302932. DOI: 10.1155/2014/302932

[16]   Swatschek D, Schatton W, Kellermann J, Müller W, Kreuter J. Marine sponge collagen: Isolation, characterization and effects on the skin parameters surface-pH, moisture and sebum. European Journal of Pharmaceutics and Biopharmaceutics. 2002;**53**:107-113

[17] Silva J, Barros A, Aroso I, Fassini D, Silva T, Reis R, Duarte A. Extraction of collagen/gelatin from the marine demosponge chondrosia reniformis (nardo, 1847) using water acidified with carbon dioxide-process optimization. Industrial & Engineering Chemistry Research. 2016;55:6922-6930. DOI: 10.1021/acs.iecr.6b00523

[18] Karim A, Bhat R. Fish gelatin: Properties, challenges, and prospects as an alternative to mammalian gelatins. Food Hydrocolloids. 2009;23:563-576. DOI: 10.1016/j.foodhyd.2008.07.002

[19] Gómez-Guillén M, Giménez B, López-Caballero M, Montero M. Functional and bioactive properties of collagen and gelatin from alternative sources: A review. Food Hydrocolloids. 2011;25(8):1813-1827. DOI: 10.1016/j.foodhyd.2011.02.007

[20] Allen F, Lanier T, Hutin H. Características de los tejidos musculares comestibles. In: Fennema O, editor. Química de los alimentos. 2nd ed. Zaragoza: Acribia; 2000. p. 1066-1071

[21] Hashim P, Ridzwan M, Bakar J, Hashim M. Collagen in food and beverage industries. International Food Research. 2015;22(1):1-8

[22] Wang L, An X, Yang F, Xin Z, Zhao L, Hu Q. Isolation and characterisation of collagens from the skin, scale and bone of deep-sea redfish (*Sebaste mentella*). Food Chemistry. 2008;108(2):616-623. DOI: 10.1016/j.foodchem.2007.11.017

[23] Pal G, Suresh P. Sustainable valorization of seafood by-products: Recovery of collagen and development of collagen-based novel functional food ingredients. Innovative Food Science & Emerging Technologies. 2016;37:201-215. DOI: 10.1016/j.ifset.2016.03.015

[24] Patino M, Neiders M, Andreana S, Noble B, Cohen R. Collagen: An overview. Implant Dentistry. 2002;11(3):280-285

[25] Ulery BD, Nair LS, Laurencin CT. Biomedical applications of biodegradable polymers. Journal of Polymer Science Part B, Polymer Physics. 2011;49(12):832-864. DOI: 10.1002/polb.22259

[26] Brett D. A review of collagen and collagen-based wound dressings. Wounds. 2008;20: 347-356

[27] Demling R. Physical and chemical factors affecting repair the epidermis in wounds. In: Rovee DT, Maibach HI, editors. The Epidermis in Wound Healing. Boca Raton: CRC Press; 2003

[28] Chattopadhyay S, Raines RT. Review collagen-based biomaterials for wound healing. Biopolymers. 2014;101:821-833. DOI: 10.1002/bip.22486

[29] Silva T, Moreira-Silva J, Marques A, Domingues A, Bayon Y, Reis R. Marine origin collagens and its potential applications. Marine Drugs. 2014;12(12):5881-5901. DOI: 10.3390/md12125881

[30] Wang H, Liang Y, Wang H, Zhang H, Wang M, Liu L. Physical-chemical properties of collagens from skin, scale, and bone of grass carp (*Ctenopharyngodon idellus*). Journal of Aquatic Food Product Technology. 2013;23(3):264-277. DOI: 10.1080/10498850.2012.713450

[31] Sampath N, Nazeer R. Characterization of acid and pepsin soluble collagen from the skin of horse mackerels (*Magalaspis cordyla*) and croaker (*Otolithes ruber*). International Journal of Food Properties. 2013;**16**(3):613-621. DOI: 10.1080/10942912.2011.557796

[32] Wang J, Pei X, Liu H, Zhou D. Extraction and characterization of acid-soluble and pepsin-soluble collagen from skin of loach (*Misgurnus anguillicaudatus*). International Journal of Biological Macromolecules. 2018;**106**:544-550. DOI: 10.1016/j.ijbiomac.2017.08.046

[33] Chen J, Li L, Yi R, Xu N, Gao R, Hong B. Extraction and characterization of acid-soluble collagen from scales and skin of tilapia (*Oreochromis niloticus*). LWT — Food Science and Technology. 2016;**66**:453-459. DOI: 10.1016/j.lwt.2015.10.070

[34] Skierka E, Sadowska M, Karwowska A. Optimization of condition for demineralization Baltic cod (*Gadus morhua*) backbone. Food Chemistry. 2007;**105**:215-218. DOI: 10.1016/j.foodchem.2007.04.001

[35] Zhang J, Duan R, Tian Y, Konno K. Characterisation of acid-soluble collagen from skin of silver carp (*Hypophthalmichthys molitrix*). Food Chemistry. 2009;**116**:318-322. DOI: 10.1016/j.foodchem.2009.02.053

[36] Wang Y, Regenstein J. Effect of EDTA, HCl, and citric acid on Ca salt removal from Asian (silver) carp scales prior to gelatin extraction. Journal of Food Science. 2009;**74**(6): C426-C431. DOI: 10.1111/j.1750-3841.2009.01202.x

[37] Nagai T, Yamashita E, Taniguchi K, Kanamori N, Suzuki N. Isolation and characterisation of collagen from the outer skin waste material of cuttlefish (*Sepia lycidas*). Food Chemistry. 2001;**72**(4):425-429. DOI: 10.1016/S0308-8146(00)00249-1

[38] Song E, Kim S, Chun T, Byun H, Lee Y. Collagen scaffolds derived from a marine source and their biocompatibility. Biomaterials. 2006;**27**(15):2951-2961. DOI: 10.1016/j.biomaterials.2006.01.015

[39] Kim H, Kim Y, Kim Y, Park H, Lee N. Effects of ultrasonic treatment on collagen extraction from skins of the sea bass *Lateolabrax japonicus*. Fisheries Science. 2012;**78**:485. DOI: 10.1007/s12562-012-0472-x

[40] Ran X, Wang L. Use of ultrasonic and pepsin treatment in tandem for collagen extraction from meat industry by-products. Journal of the Science of Food and Agriculture. 2014;**94**(3):585-590. DOI: 10.1002/jsfa.6299

[41] Yu F, Zong C, Jin S, Zheng J, Chen N, Huang J, Ding G. Optimization of extraction conditions and characterization of pepsin-solubilised collagen from skin of giant croaker (*Nibea japonica*). Marine Drugs. 2018:16-29. DOI: 10.3390/md16010029

[42] Huang C, Kuo J, Wu S, Tsai H. Isolation and characterization of fish scale collagen from tilapia (*Oreochromis sp.*) by a novel extrusion-hydro-extraction process. Food Chemistry. 2016;**190**:997-1006. DOI: 10.1016/j.foodchem.2015.06.066

[43] Neklyudov A. Nutritive fibers of animal origin: Collagen and its fractions as essential components of new and useful food products. Applied Biochemistry and Microbiology. 2003;**39**(3):229-238. DOI: 0.1023/A:1023589624514

[44] Bilek S, Bayram S. Fruit juice drink production containing hydrolyzed collagen. Journal of Functional Foods. 2015;**14**:562-569. DOI: 10.1016/j.jff.2015.02.024

[45] Helfrich Y, Sachs D, Voorhees J. Overview of skin aging and photoaging. Dermatology Nursing. 2008;**20**:177

[46] Albu M, Titorencu I, Ghica M. Collagen-based drug delivery systems for tissue engineering. In: Pignatello Rosario, editor. Biomaterials Applications for Nanomedicine. Croatia: InTech; 2001. p. 333-358. DOI: 10.5772/22981

[47] Bayón B, Berti I, Gagneten A, Castro G. Biopolymers from Wastes to High-Value Products in Biomedicine. In: Singhania R, Agarwal R, Kumar R, Sukumaran R, editors. Waste to Wealth. Energy, Environment, and Sustainability. Singapore: Springer; 2018. p. 1-44. DOI: 10.1007/978-981-10-7431-8_1

[48] Chandika P, Ko S, Oh G, Heo S, Nguyen V, Jeon Y, Chang W. Fish collagen/alginate/chitooligosaccharides integrated scaffold for skin tissue regeneration application. International Journal of Biological Macromolecules. 2015;**81**:504-513. DOI: 10.1016/j.ijbiomac.2015.08.038

[49] Cheng X, Shao Z, Li C, Yu L, Raja M, Liu C. Isolation, characterization and evaluation of collagen from jellyfish rhopilema esculentum kishinouye for use in hemostatic applications. PLoS One. 2017;**12**(1):e0169731. DOI: 10.1371/journal. pone.0169731

[50] Ullah S, Zainol I, Chowdhury S, Fauzi M. Development of various composition multicomponent chitosan/fish collagen/glycerin 3D porous scaffolds: Effect on morphology, mechanical strength, biostability and cytocompatibility. International Journal of Biological Macromolecules. 2018;**111**:158-168. DOI: 10.1016/j.ijbiomac.2017.12.136

[51] Pallela R, Venkatesan J, Janapala V. Biophysicochemical evaluation of chitosan-hydroxyapatite-marine sponge collagen composite for bone tissue engineering. Journal of Biomedical Materials Research. Part A. 2012;**100**(2):486-495. DOI: 10.1002/jbm.a.3329 x

[52] Veeruraj A, Arumugam M, Ajithkumar T, Balasubramanian T. Isolation and characterization of drug delivering potential of type-I collagen from eel fish Evenchelys macrura. Journal of Materials Science. Materials in Medicine. 2012;**23**(7):1729-1738. DOI: 10.1007/s10856-012-4650-2

[53] Tang L, Chen S, Su W, Weng W, Osako K, Tanaka M. Physicochemical properties and film-forming ability of fish skin collagen extracted from different freshwater species. Process Biochemistry. 2015;**50**:148-155. DOI: 10.1016/j.procbio.2014.10.015

[54] Perumal R, Perumal S, Thangam R, Gopinath A, Ramadass S, Madhan B, Sivasubramanian S. Collagen-fucoidan blend film with the potential to induce fibroblast proliferation for regenerative applications. International Journal of Biological Macromolecules. 2018;**106**:1032-1040. DOI: 10.1016/j.ijbiomac.2017.08.111

[55] Liu T, Shi L, Gu Z, Dan W, Dan N. A novel combined polyphenol-aldehyde crosslinking of collagen film – Applications in biomedical materials. International Journal of Biological Macromolecules. 2017;**101**:889-895. DOI: 10.1016/j.ijbiomac.2017.03.166

[56] Raiskup F, Theuring A, Pillunat L, Spoerl E. Corneal collagen crosslinking with ribo-flavin and ultraviolet – A light in progressive keratoconus: Ten-year results. Journal of Cataract and Refractive Surgery. 2015;**41**(1):41-46. DOI: 10.1016/j.jcrs.2014.09.033

[57] Long Y, Zhao X, Liu S, Chen M, Liu B, Ge J, et al. Collagen-hydroxypropyl methylcellu-lose membranes for corneal regeneration. ACS Omega. 2018;**3**:1269-1275. DOI: 10.1021/acsomega.7b01511

[58] Pallela R, Venkatesan J, Bhatnagar I, Shim Y, Kim S. Applications of marine collagen-based scaffolds in bone tissue engineering. In: Kim S-K, editor. Marine Biomaterials: Characterization, Isolation and Applications. Boca Raton: Taylor & Francis; 2013. pp. 519-552. DOI: 10.1201/b14723-30

[59] Dickson G, Buchanan F, Marsh D, Harkin-Jones E, Little U, McCaigue M. Orthopaedic tissue engineering and bone regeneration. Technology and Health Care. 2007;**15**:57-67

[60] Aravamudhan A, Ramos D, Nip J, Kalajzic I, Kumbar S. Micro-nanostructures of cellulose-collagen for critical sized bone defect healing. Macromolecular Bioscience. 2017;**18**:1700263. DOI: 10.1002/mabi.201700263

[61] Wahl D, Czernuszka J. Collagen-hydroxyapatite composites for hard tissue repair. European Cells & Materials. 2006;**28, 11**:43-56

[62] Murphy C, Haugh M, O'Brien F. The effect of mean pore size on cell attachment, prolif-eration and migration in collagen-glycosaminoglycan scaffolds for bone tissue engineer-ing. Biomaterials. 2010;**31**(3):461-466. DOI: 10.1016/j.biomaterials.2009.09.063

[63] Calejo M, Almeida A, Fernandes A. Exploring a new jellyfish collagen in the production of microparticles for protein delivery. Journal of Microencapsulation. 2012;**29**:520-531. DOI: 10.3109/02652048.2012.665089

[64] Dinescu S, Gălățeanu B, Albu M, Lungu A, Radu E, Hermenean A, Costache M. Biocompatibility assessment of novel collagen-sericin scaffolds improved with hyal-uronic acid and chondroitin sulfate for cartilage regeneration. BioMed Research International. 2013;**2013**:598056. DOI: 10.1155/2013/598056

[65] Langasco R, Cadeddu B, Formato M, Lepedda A, Cossu M, Giunchedi P, Gavini E. Natural collagenic skeleton of marine sponges in pharmaceutics: Innovative biomaterial for topical drug delivery. Materials Science & Engineering. C, Materials for Biological Applications. 2017;**1**:710-720. DOI: 10.1016/j.msec.2016.09.041

[66] Pérez J, Muñoz E, Huaquín C, Nirchio M. Riesgos de la introducción de tilapias (*Oreochromis sp.*) (*Perciformes: Cichlidae*) en ecosistemas acuáticos de Chile. Revista Chilena de Historia Natural. 2004;**77**(1):195-199. DOI: 10.4067/S0716-078X2004000100015

[67] Fontaínhas-Fernandes A, Russell-Pinto F, Gomes E, Reis-Henriques MA, Coimbra J. The effect of dietary sodium chloride on some osmoregulatory parameters of the teleost, *Oreochromis niloticus*, after transfer from freshwater to seawater. Fish Physiology and Biochemistry. 2001;**23**:307-316. DOI: 10.1023/A:10111568

[68] Jiménez, R. Enfermedades de tilapia en cultivo. Universidad de Guayaquil. 1 Ed. Facultad de ciencias naturales. Guayaqui, Ecuador. 2007:110 p

[69] Singh P, Benjakul S, Maqsood S, Kishimura H. Isolation and characterisation of collagen extracted from the skin of striped catfish (*Pangasianodon hypophthalmus*). Food Chemistry. 2010;**124**:97-105. DOI: 10.1016/j.foodchem.2010.05.111

[70] AOAC official methods of analysis 16th ed. Washington, DC: Association of Official Analytical. 1995

[71] Tamilmozhi S, Veeruraj A, Arumugam M. Isolation and characterization of acid and pepsin-solubilized collagen from the skin of sailfish (*Istiophorus platypterus*). Food Research International. 2013;**54**(2):1499-1505. DOI: 10.1007/s10668-014-9538-5

[72] Hickman D, Sims T, Miles C, Bailey A, De Mari M, Koopmans M. Isinglass/collagen: Denaturation and functionality. Journal of Biotechnology. 2000;**79**(3):245-257. DOI: 10.1016/S0168-1656(00)00241-8

[73] Sun L, Hou H, Li B, Zhang Y. Characterization of acid-and pepsin-soluble collagen extracted from the skin of Nile tilapia (*Oreochromis niloticus*). International Journal of Biological Macromolecules. 2017;**99**:8-14. DOI: 10.1016/j.ijbiomac.2017.02.057

[74] Okazaki E, Osako K. Isolation and characterization of acid-soluble collagen from the scales of marine fishes from Japan and Vietnam. Food Chemistry. 2014;**149**:264-270. DOI: 10.1016/j.foodchem.2013.10.094

[75] Hsieh C, Shiau C, Su Y, Liu Y, Huang Y. Isolation and characterization of collagens from the skin of giant grouper (*Epinephelus lanceolatus*). Journal of Aquatic Food Product Technology. 2016;**25**:93-104. DOI: 10.1080/10498850.2013.828145

[76] Fu W, Wang Y, Zheng B, Liao M, Zhang W. Isolation and characterization of pepsin-soluble collagen from the skin of Peru squid (*Dosidicus gigas*). Journal of Aquatic Food Product Technology. 2012;**22**:270-280. DOI: 10.1080/10498850.2011.646002

[77] Montero P, Jimennez-Colmenero F, Borderias J. Effect of pH and the presence of NaCl on some hydration properties of collagenous material from trout (*Salmo irideus* Gibb) muscle and skin. Journal of Science and Food Agriculture. 1991;**54**:137-146. DOI: 10.1002/jsfa.2740540115

[78] Bradford MM. A rapid and sensitive method for the quantitation of microgram quantities of protein utilizing the principle of protein-dye binding. Analytical biochemistry. 1976;**72**(1-2):248-254

[79] Woo J, Yu S, Cho S, Lee Y, Kim S. Extraction optimization and properties of collagen from yellowfin tuna (*Thunnus albacores*) dorsal skin. Food Hydrocolloids. 2007;**22**(5):879-887. DOI: 10.1016/j.foodhyd.2007.04.015

[80] Bae I, Osatomi K, Yoshida A, Osako K, Yamaguchi A, Hara K. Biochemical properties of acid-soluble collagens extracted from the skins of underutilised fishes. Food Chemistry. 2008;**108**:49-54. DOI: 10.1016/j.foodchem.2007.10.039